PAPERS IN MATHEMATICS

ANNALS OF THE NEW YORK ACADEMY OF SCIENCES

Volume 321

PAPERS IN MATHEMATICS

Edited by Paul R. Meyer

The New York Academy of Sciences
New York, New York
1979

Library of Congress Cataloging in Publication Data

Library of Congress Cataloging in Publication Data
Main entry under title:

Papers in mathematics.

(Annals of the New York Academy of Sciences; v. 321)
"Collection of some of the papers presented at the monthly meetings of the Mathematics Section of the New York Academy of Sciences."
1. Mathematics—Addresses, essay, lectures.
I. Meyer, Paul Richard, 1930– II. New York Academy of Sciences. Mathematics Section. III. Series: New York Academy of Sciences. Annals; v. 321.
Q11.N5 vol. 321 [QA7] 508'.1s [510] 79-13706

SP

Printed in the United States of America

ISBN 0-89766-026-9

ANNALS OF THE NEW YORK ACADEMY OF SCIENCES
VOLUME 321
May 14, 1979

PAPERS IN MATHEMATICS*

Editor
PAUL R. MEYER

CONTENTS

*This volume is a collection of various papers presented at the monthly meetings of the Mathematics Section of The New York Academy of Sciences.

INTRODUCTION

Paul R. Meyer

Department of Mathematics
Herbert H. Lehman College
City University of New York
Bronx, New York, 10468

This is a collection of some of the papers presented at the monthly meetings of the Mathematics Section of The New York Academy of Sciences during the last few years. It is appropriate to say a few words here about the philosophy behind the choices of speakers and topics for these section meetings.

A researcher in science today is under tremendous pressure to specialize. One way to seek new and, one hopes, publishable results is to narrow one's focus. Unfortunately, the old cliché about "learning more and more about less and less, until one knows everything about nothing" comes uncomfortably close to describing some of what has been done recently in the name of mathematical research.

Of course, researchers must specialize. And, of course, they need a colloquium where they can discuss their findings with others in their field. Fortunately, the New York City area is rich in such specialized colloquia. We at The New York Academy of Sciences have chosen a different course: to try to counter the tendency toward overspecialization. We offer a generalized colloquium in which significant mathematics of general interest is presented in a manner intelligible to a mathematician who is not a specialist in the topic under discussion.

At a typical meeting, about half of those present are college mathematics faculty members, with the rest of the audience made up mainly of people involved in applied mathematics and of mathematics students. A majority of the mathematicians have an active research interest of their own. With this audience in mind, we have suggested to our speakers that they pitch their expositions at about the level of the American Mathematical Monthly.

The papers presented here range widely, both in content and in organization. They provide a representative sample of the flavor of what has been done in the Section in the last few years.

NONASSOCIATIVE MULTIPLICATION

Michael Aissen

Department of Mathematics
Rutgers University
Newark, New Jersey 07102

Brian Shay

Department of Mathematics
Hunter College
City University of New York
New York, New York 10021

It is the purpose of this note to surprise the reader with an analysis of familiar and central questions in the study of nonassociative binary systems with technique restricted to the concatenation of sequences of natural numbers and application of the successor function to the natural numbers. The treatment will be naive, and roughly correspond to the presentation of these ideas to the members of the New York Academy of Sciences by the second author in the spring of 1977. Proofs, an elaboration of this material, and its relationship to the work of other authors can be found in Aissen and Shay.[1]

It is our ultimate intention to gain a better understanding of the influence of laws which are strictly weaker than the generalized associative law. The simplest example of such a law appears in close relation to abelian groups, and is the set of logical consequences of the formula: $((x - (y - z)) - w) = (x - (y - (z - w)))$, just as the generalized associative law is the set of logical consequences of the formula: $((x*y)*z) = (x*(y*z))$.

Put another way, the generalized associative law is this: There are two formally distinct ways to multiply three elements (in a given order) in a binary system. There are $C_n = (1/(2n - 1)\binom{2n-1}{n}$ formally distinct ways to multiple n elements (in a given order). If the two formally distinct ways have the same value for any three arguments in a given binary system, then the C_n formally distinct ways have the same value for any n arguments in that binary system.

The "subassociative law of subtraction" is this: If in a given binary system with set X and operation $*$, for all a, b, c, d in X, the values $a*(b*(c*d))$ and $(a*(b*c))*d$ are the same, then the C_n formally distinct ways of multiplying n arguments have at most 2^{n-2} values if n is greater than 1. In fact, the following is true, but tedious to prove: if in a binary system there are at most four ways to multiply four elements, then there are at most 2^{n-2} ways to multiply n elements in that system. It happens that 2^{n-2} is not always the best "universal" upper bound. We know the best "universal" upper bounds, but the proofs are extremely complicated in some cases. It is not feasible to look for anything but asymptotic results of this kind when identifications occur only in the multiplication of more than four elements.

The least complicated set with multiplication is the free binary system on one generator. This is probably familiar to the reader in one or another guise. To emphasize a universal property, we shall use the following notation: the generator will

0077-8923/79/0321-0003 $01.75/1 © 1979, NYAS

be donated $-$; if u and v are elements of the set underlying the system, $(u*v)$ will denote their product; the set itself will be denoted F. Each element of F other than $-$ can be factored uniquely into a product of two elements. As a consequence of the unique factorization, F has a natural gradation by the natural numbers. The elements of degree 1, 2, 3, and 4 (i.e., the elements of F_1, F_2, F_3, F_4) are: $-$;$(-*-)$; $((-*-)*-)$, $(*(-*(-*-))$; $(((-*-)*-)*-)$, $((-*-)*(-*-))$, $((-*(-*-))*-)$, $(-*((-*-)*-))$, $(-*(-*(-*-)))$. The number of elements of F_n is equal to the nth Catalan number C_n.

The importance of the binary system $(F,)*))$ is this: there is a natural interpretation of any element u of F as set map from the n-fold Cartesian product of Y into Y, if n is the degree of u and $(Y, \circ)$ is a binary system. For example, $((-*-)*-)$ is to be interpreted as a map from $Y \times Y \times Y$ to Y, with value $((y_1 \circ y_2) \circ y_3)$ on the triple (y_1, y_2, y_3). Consequently, an element of F will be called a formal product. It takes little imagination to generalize this interpretation as follows: suppose $u = (u_1, u_2, \ldots, u_k)$ belongs to the k-fold Cartesian product of F, and n_i is the degree of u_i for $i = 1, 2, \ldots, k$. Then u has a natural interpretation as a set map from the n-fold to the k-fold Cartesian product of Y, if n is the sum of $n_1, n_2, \ldots, n_k$, and $(Y,\circ)$ is a binary system. For example, $((-*-),((-*-)*-))$ has value $((y_1 \circ y_2),((y_3 \circ y_4) \circ y_5))$ at $(y_1, y_2, y_3, y_4, y_5)$. In particular, if $(Y,\circ)$ is $(F,(*))$ and y is $(-,-,\ldots,-)$, then $u(y)$ is u. If y is any n-tuple of elements of F, $u(y)$ is said to be the substitution of y in u. A partial order on F is defined by demanding that v be less than any substitution in v if v is a formal product. This partial order induces the structure of a lattice on F, to be called the substitution lattice.

In the notation of the previous paragraph, the k-tuple of formal products u will be called a generalized formal product of bidegree (n, k). The bigraded set of generalized formal products will be denoted A. The set A has several binary operations or partial operations of interest:

(i) Cartesian product: $A_{n,k} \times A_{n',k'}$ to $A_{n+n',k+k'}$
(ii) composition: $A_{p,n} \times A_{n,k}$ to $A_{p,k}$
(iii) $(*)$: $A_{n,1} \times A_{n',1}$ to $A_{n+n',1}$

for all admissible n,n', k, k', p (first element of the bidegree must be greater than or equal to the second).

We shall describe a many-to-one mapping from the set of finite sequences of natural numbers to the set of generalized formal products. The image will not be all generalized formal products, but only those not of the form $(u_1, u_2, \ldots, u_k, -)$. In short, the subset of A which we do not parametrize is the set $A \times -$, which has no structure of any interest that does not already appear in the subset of A which we do parametrize. Another point of view is this: for each pair of natural numbers (n, k), with n greater than or equal to k, there is a trivial inclusion, $f_{n,k}$ of $A_{n,k}$ into $A_{n+1,k+1}$ consisting of "padding with a blank of the right." The object we should be studying is the directed system of sets and maps generated by the A's and f's.

We shall call finite sequences in N, the natural numbers, strings; let S denote the set of strings. This set is to be associated below with the set of generalized formal products, and the simplest structure it will display is a compartmentalization into bigraded components. If $a = [a_1, a_2, \ldots, a_j]$ is a string, let $L(a) = j$, $R(a) = 1 +$

$\text{Max}_i\ (a_i - i)$. Since a_1 is greater than 0, $R(a)$ is greater than 0. The string is in component $S_{n,k}$ if $L(a) + R(a) = n$ and $R(a) = k$. We use the convention that the empty string has bidegree (1, 1) and is alone in its component. The string 6, 3, 2, 1, 2, 1, has bidegree (12, 6), for example.

If $R(a) = 1$, we have at hand a version of a formal product, not only a generalized formal product. This condition is clearly equivalent to: a_1 is less than or equal to 1, a_2 is less than or equal to 2, . . . , a_j is less than or equal to j.

Now that a decision has been made about bidegree, we are faced with interpreting an element of $S_{n,k}$ as a map from the n-fold to the k-fold Cartesian product of the set of any binary system. As the example given earlier suggests, it suffices to give such an interpretation for the special case $(F,(*))$, and it is only necessary to define the value of the map on the n-tuple $(-,-,\ldots,-)$.

We proceed with an algorithm:

(i) define $b_i = a_{j-i+1}$ (i.e., reverse the string);
(ii) define $y_{0i} = -$, $i = 1, 2, \ldots, n$;
(iii) if y_{ri} is defined, $i = 1, 2, \ldots, n - r$, define y_{r+1i} as

y_{ri}, if i is less than b_r
$(y_{ri} * y_{ri+1})$, if i is b_r
Y_{ri+1}, if i is greater than b_r;

(At this point, some readers will be thinking of face operators and semi-simplicial complexes)

(iv) stop if $r = L(a) = n - k = j$.

The result is an element of A to which we map the string a. We shall denote this map from S to A by f. As mentioned above, the image of f is $A/(A \times -)$.

The following example will illustrate the algorithm. Let $a = [6, 3, 2, 1, 2, 1]$:

start: $(-,-,-,-,-,-,-,-,-,-,-,-)$;
next: $((-*-),-,-,-,-,-,-,-,-,-,-)$; $(b_1 = 1)$
next: $((-*-),(-*-),-,-,-,-,-,-,-,-)$; $(b_2 = 2)$
next: $(((-*-)*(-*-)),-,-,-,-,-,-,-,-)$; $(b_3 = 1)$
next: $(((-*-)*(-*-)),(-*-),-,-,-,-,-,-)$; $(b_4 = 2)$
next: $(((-*-)*(-*-)),(-*-),(-*-),-.-,-,-)$; $(b_5 = 3)$
next: $(((-*-)*(-*-)),(-*-),(-)*\ -),-,-,(-*-))$; $(b_6 = 6)$
stop.

The last formal product in the k-tuple is decomposable. This is a characteristic property, with the following exception: let $a = [\ \ ]$:

start: $(-)$;
stop.

We now consider the natural question: Which sequences are mapped to the same generalized formal product by the algorithm? To fix notation, let us say that $a \doteq b$ if $f(a) = f(b)$. It is quite clear that $\doteq$ is an equivalence relation. As a matter of fact, it is a congruence relation with respect to concatenation of strings, the familiar operation, indicated by juxtaposition, and defined as follows: $[a_1, a_2, \ldots, a_{bg}][b_1, b_2, \ldots, b_h] = [a_1, \ldots, a_g, b_1, \ldots b_h]$.

THEOREM 1 (i) If a, b, c, and d are strings and $a = b$, then $cad \doteq cbd$;
(ii) If i and j are natural numbers and j is greater than i, $[j, i] \doteq [i, j+1]$;
(iii) $\doteq$ is the smallest equivalence relation satisfying (i) and (ii).

The motivation for enlarging our discussion from formal products to generalized formal products is this theorem. Clearly the generators of the congruence relation, as described in (ii), do not represent equivalences between strings which represent formal products, so there is no analog of THEOREM 1 that says something as interesting about just those strings.

Our immediate goal is to get an interesting model for the binary system $(F,(*))$, and not only the set F. A binary operation has come into play in THEOREM 1, but it is an associative operation, and cannot be a version of $(*)$ when restricted to suitable arguments.

For the simplest presentation, we must also consider a unary operation on strings. The operation, denoted $+$, simply adds 1 to each natural number in a string. If $a = [a_1, a_2, \ldots, a_j]$, then $a^+ = [a_1 + 1, a_2 + 1, \ldots, a_j + 1]$. A nonassociative operation on S, denoted #, is defined as follows: if a and b are strings, $a\#b = [1]b^+a$.

THEOREM 2. If $a \doteq b$ and $c \doteq d$, then $a\#c \doteq b\#d$.

In short, although, by THEOREM 1 (iii), we may think of the equivalence relation $\doteq$ as being defined as a congruence with respect to concatenation, # is such a "slight modification" of concatenation that it induces a binary operation $\dot{\#}$ on the set of $\doteq$ equivalence classes. Consider the following example: $f([3, 2]\#[1, 3]) = f([3, 2]) \;\dot{\#}\; f([1, 3]) = (-, (-*-), (-*-)) \;\dot{\#}\; ((-*-), (-*-)) = ((-*((-*-)*(-*-))), (-*-)) = f([1, 2, 4, 3, 2])$. This does *not* look familiar, but the justification for the definition of # is this:

THEOREM 3. If a and b are strings and $f(a)$ and $f(b)$ are formal products, then $f(a\#b) = f(a) \;\dot{\#}\; f(b) = (f(a)*f(b))$.

Thus, we have completed our model for $(F,(*))$. It will be easy to work with if good representatives for the $\doteq$ equivalence classes can be chosen and if it is easy to determine whether or not a string is equivalent to a particular one of these good representatives.

The first of these requirements is met by the following:

THEOREM 4. In each $\doteq$ equivalence class, there is a unique nondecreasing sequence of natural numbers.

Still more structure will be imposed on S to ensure that it cannot be difficult to determine *which* nondecreasing string is equivalent to a given string.

Let S be given a partial ordering as follows:

(i) if i and j are natural numbers and i is less than j, $[j, i]$ is less than $[i, j+1]$;

(ii) if a, b, and c are strings, a is less than b, and b is less than c, then a is less than c;
(iii) if a, b, c, and d are strings and a is less than b, then cad is less than cbd;
(iv) if a, b, and c are strings and ac is less than b, then a is less than b.

The reader should notice that (i), (ii), and (iii) would be concerned with comparisons within an equivalence class without the presence of (iv).

THEOREM 5. (i) The partial order on S induces the structure of a lattice on S;
(ii) the restriction of the partial order to each equivalence class induces the structure of a lattice on that class; moreover, the supremum will be the unique nondecreasing string in the class;
(iii) if a and b are strings representing formal products, and a is less than b, then $f(a)$ is less than $f(b)$ in the substitution lattice; moreover, the restriction on the partial order to these strings induces the structure of a lattice;
(iv) if u and v are comparable formal products in the substitution lattice, and u is less than v, then there are strings a, b, and c satisfying $ac = $ b, $f(a) = u$, $f(b) = v$.

The most interesting detail of THEOREM 5 is the relationship between concatenation of strings and substitution of formal products. However, we must direct our attention to the following consequence. Suppose we wish to find the nondecreasing string equivalent to a given string. If any adjacent pair of natural numbers in the given string is decreasing, the pair can be switched and 1 added to the larger of the two. The resulting string will be in the same equivalence class and closer to the supremum of the lattice. Since the class is finite, *any* rule for selecting the pair will work.

The following example should suffice to clarify this point:

$$[6, 3, 2, 1, 2, 1] \doteq [3, 7, 2, 1, 2, 1] \doteq [3, 2, 8, 1, 2, 1] \doteq [3, 2, 1, 9, 2, 1] \doteq [3, 2, 1, 2, 10, 1] \doteq [3, 2, 1, 2, 1, 11] \doteq [3, 2, 1, 1, 3, 11] \doteq [2, 4, 1, 1, 3, 11] \doteq [2, 1, 5, 1, 3, 11] \doteq \ldots \doteq [1, 1, 3, 5, 7, 11]$$

A basic result is the identification of the nth Catalan number with the size of the graded component F_n. Because we have identified F_n with nondecreasing sequences of n-1 natural numbers less than or equal to the sequence $[1, 2, \ldots, n\text{-}1]$, this result can be verified by counting such sequences (this is not new fact.)

Certain congruences with respect to the operations (Cartesian product, composition, ($*$)) on the set of generalized formal products seem to deserve the designation "subassociative laws." They can be shown to correspond to congruences on strings with respect to the concatenation and + operations which contain $\doteq$. The simplest of these is the generalized associative law. One might expect a generator of the associated congruence relation on strings to be the equivalence of [1, 1] (representing $((-*-)*-)$) and [1, 2] (representing $(-*(-*-))$). The smallest congruence with respect to concatenation and + containing ([1, 1], [1, 2]) and $\doteq$ is the smallest congruence with respect to concatenation containing $([j, i], [i, j+1])$, if i and j are natural numbers and i is less than *or equal to* j. Let us denote this congruence by e.

THEOREM 6 (generalized associative law). For each natural number n, $S_{n,1}$ is an e-equivalence class.

Proof: This is clear if $n = 1$. Suppose it to be true if n is less than k. The standard representative will be $[1, 1, \ldots, 1]$. In $S_{k,1}$, $a = [a_1, a_2, \ldots, a_{k-1}, a_k] \doteq [a_1, a_2, \ldots, a_{k-1}][a_k]$ e $[1, 1, \ldots, 1][a_k]$ e $[1, 1, \ldots, 1, a_k]$ e $[1, 1, \ldots, a_k - 1, 1]$ e $[1, 1, \ldots, a_k\text{-}1][1]$ e $[1, 1, \ldots, 1][1]$ e $[1, 1, \ldots, 1, 1]$.

REFERENCE

1. AISSEN, M. & B. SHAY. 1977. Varieties of binary systems. J. Comb. Theory **22**(1): 69–82.

DECIDING SOME UNDECIDABLE TOPOLOGICAL STATEMENTS*

W. W. Comfort

Department of Mathematics
Wesleyan University
Middletown, Connecticut 06457

§1. Introduction

I appreciate the kind introductory remarks and I am grateful also to Paul Meyer and the Organizing Committee for the invitation to be here; it is a pleasure and an honor to address the Academy.

I want to describe some axioms that have made an appearance in the literature and some of the problems they have helped to solve or settle. Since I am principally a topologist, and not a logician or a set theorist, my chief hope is to interest or amuse you with the topological considerations which arise—the axioms are a vehicle for approaching the topological problems. I should warn you at the outset that, although the axioms we shall consider differ from one another and serve different (sometimes even competing) purposes, I am not building up to a final recommendation which serves the human condition optimally. As with many decisions we make in our lives, different choices are best in different respects. There is room for personal taste and, in any event, no social or professional stigma accrues to the mathematician who, finding that an axiom formerly embraced has lost its appeal, discards it in favor of another.

The scientific work of most modern mathematicians is not shaped, or even affected, by any coherent philosophy concerning the world we inhabit or the rules of play that are reasonable and acceptable. We are willing to adopt new axioms as the occasion demands, with no more sense of loss for the old ones than we feel in the morning about the socks we took off the night before.

There are, of course, prominent mathematicians who care deeply about axiom schemes and permissible rules of procedure. One thinks immediately in this connection of Errett Bishop,[3] who is active today, and of L. E. J. Brouwer (1881–1966), whose well-known contributions to the foundations of algebraic topology[10] were achieved in the context of severe doubts about the legitimacy of several generally accepted practices, including argument by appeal to the law of the excluded middle.[8] Brouwer's misgivings on these matters appeared, though in tentative and unpolished form, as early in his career as his doctoral dissertation[7]; according to Freudenthal and Heyting,[37] he was to assert some years later that: "... in his topological work he had tried to use only methods which he expected could be made constructive." A vigorous appreciation of Brouwer and his historical importance is given by Kreisel.[57]

It should be remarked that the modern set theorist or logician is, in his selection of axioms to work with, not guided exclusively by his intuition as to what is or ought to be true of our (physical) world. Just as physicists have been forced in recent decades, over

*Partially supported by National Science Foundation grant NSF-MCS 76-05821.

0077–8923/79/0321–0009 $01.75/1 © 1979, NYAS

and over again, to reshape their thinking as to the principles which govern the behavior of the atom and the cosmos, so have mathematicians had to accommodate to new, unsettling developments.

Not every question which can be asked can be answered. In seeking interesting axioms which determine or settle difficult questions, today's set theorist is not restricted to principles which are intuitively or obviously true. It appears that, having been forced to distrust our intuition, we have abandoned it for the time being. An interesting axiom or principle which settles or determines a nontrivial question becomes, by virtue of that power, suitable for use and further investigation. Let us try first to understand the necessity for new axioms.

§2. True and Undecidable?

I begin by quoting a sentence spoken to me about two decades ago by a mathematician I admire and respect. I am quoting from memory, but the sentence struck me at the time as so remarkable, so implausible, so bizarre, so ridiculous, that I think I caught the words verbatim: "I recognize that Fermat's Last Theorem may be undecidable; in that case of course it's true." At the time of this conversation I could not have given a careful, correct definition of either "undecidable" or "true," but I certainly felt I knew enough about each to be certain that no statement could be both. If we have determined enough about a statement to have decided that it is true, surely the statement cannot be undecidable. Let us try to see why my confusion was unnecessary and why the sentence quoted is quite reasonable.

Historically, we find numerous examples of the following phenomenon. Some mathematician—say Euclid[34] as a compiler, or Peano,[77] or Zermelo,[104] or Fraenkel[35,36]—draws up a short list of axioms or postulates intended to describe the real world, or the arithmetic properties of the integers, or set theory, or some other substantial part of Mathematics. Initially in each case, the hope, and even the expectation, has been that the axioms proposed will prove to be categorical—that is, that any two systems or models satisfying the axioms will be isomorphic in an appropriate sense.

We understand now, thanks to the work of Gödel,[40] that such a project cannot succeed. Gödel's Incompleteness Theorem asserts, roughly, that any recursive axiomatization of the integers, and, more generally, any consistent formal axiomatized system containing both the integers and the primitive recursive functions, is incomplete in the sense that there is a statement S such that neither S nor its negation $\neg S$ is derivable from the axioms. In effect, Gödel lists the countably many arguments which might serve to prove or to refute a statement meaningful in the system in question and then formulates a statement S not settled by any of the arguments; the procedure for producing S is similar to, though technically more complicated than, the diagonalization process introduced by Cantor[11] to show that the set of real numbers is uncountable.

Let us return to the sentence I quoted asserting that certain natural statements, among them perhaps Fermat's Last Theorem, might be undecidable and, simultaneously, true. (The Theorem itself, you will recall, is the statement that if a, b, c, and n are positive integers with $n > 2$, then $a^n + b^n \neq c^n$.) Let us suppose for the moment, just for the sake of argument, that Fermat's Last Theorem is in fact undecidable. That means that on the basis of the axioms we are using and the commonly accepted rules of inference it is impossible to give a proof and it is impossible to define a counterexample.

In any specific model, however, Fermat's Last Theorem is either true or false. In a specific model we may be able, using tricks or techniques not applicable (perhaps not even meaningful) in some other models, to determine which. By good fortune, our own usual model **Z** of the integers, consisting of $\{0, \pm 1, \pm 2, \ldots\}$, together with the arithmetic operations familiar to us, has the property that we can write a computer program which, if there is a 4-tuple $k = \langle a, b, c, n\rangle$ of positive integers with $n > 2$ and $a^n + b^n = c^n$, will find it. (To write such a program is a simple assignment within the scope of a beginning student. One may, for example, having chosen a language of communication appropriate to the computer at hand, define $f(k) = a + b + c + n$ for k as above, and then arrange, for $m = 5, 6, \ldots$, that the computer list all 4-tuples $k = \langle a, b, c, n\rangle$ such that $f(k) = m$ and check, in each case, whether or not $a^n + b^n = c^n$; we need not demand a printout of each instance of failure, but we should make certain that the computer informs us when and if it finds k for which the equality holds.) Now we reason as follows: If there is a counterexample to Fermat's Last Theorem in (our model of) the integers, we will find it; hence if Fermat's Last Theorem is undecidable on the basis of the axioms of set theory, then that theorem is true (in our model). We see, then, how it might occur that a statement is both undecidable and true.

For the sake of emphasis I note explicitly that it is not known at this time whether Fermat's Last Theorem is decidable on the basis of the axioms of Zermelo-Fraenkel set theory. In any event there are concrete (combinatorial) statements about the integers which, though true, cannot be settled in Zermelo-Fraenkel set theory. According to the recent solution of Matijasevič to the tenth problem of Hilbert,[47] there is a polynomial $P(x_0, \ldots, x_{13})$ with integral coefficients such that

$$(*) \text{ if } x_0, \ldots, x_{13} \text{ are integers then } P(x_0, \ldots, x_{13}) \neq 0$$

but $(*)$ cannot be proved in Zermelo-Fraenkel set theory. (The history of Hilbert's tenth problem, and of the contributions toward its solution by M. Davis, J. Robinson, Yu. Matijasiveč, and others are available in Davis[27,28] and, in detail accessible to the layman, in Davis.[26]) The polynomial P is a clever coding of an undecidable statement, not studied earlier in its own right and of no number-theoretic interest as a Diophantine equation. In contrast Paris and Harrington[76] have shown that a certain very natural result in finitary combinatorics—a simple extension of the Ramsey Theorem—is true but unprovable on the basis of the usual first-order axioms of Peano arithmetic.

§3. Euclid's Fifth Postulate

In the hope that it will clarify the point being made here I want to describe another instance of this phenomenon—this one familiar to us all. I have in mind the status of Euclid's fifth postulate relative to the other four. To simplify slightly, let us consider this postulate in the form adopted by the Euclidean commentator Proclus and, some thirteen centuries later, by the Scottish physicist, John Playfair: Through a given point can be drawn exactly one line parallel to a given line. We understand today, thanks to the work of Lobachevsky[66,67] and Bolyai,[5] that this postulate cannot be proved from the others because it is in fact false in certain systems which are models of the other four; these are the so-called non-Euclidean geometries (e.g., the hyperbolic geometry of Klein[54,55]). Whether or not the world we live in is Euclidean (it is not quite so, according to Einstein, since space is slightly curved), you and I can probably agree on a mutually satisfactory definition of a point and line. To us a point is a real ordered triple $P = \langle x, y, z\rangle$ and a line is, for some point $\bar{P} = \langle \bar{x}, \bar{y}, \bar{z}\rangle\rangle$ and for some real ordered triple $\langle a, b, c\rangle$, not all zero, the set L of points $\langle \bar{x} + ta, \bar{y} + tb, \bar{z} + tc\rangle$, with $t \in \mathbf{R}$. We can define the concept of parallelism between lines and, given a point $P_0 = \langle x_0, y_0, z_0\rangle$, we can find and prove the uniqueness of that line through P_0 and parallel to L. It is, in fact, the line given by $\langle x_0 + ta, y_0 + tb, z_0 + tc\rangle$.

Euclid's fifth postulate is, then, undecidable and true. More fully: It is undecidable from the other four, and it is true in our everyday model of Euclidean 3-space.

I note in passing that the statement given above of Gödel's Incompleteness Theorem was highly informal and itself quite incomplete. This is not the time for a careful and accurate statement, but it is worth pointing out that the requirement that the axioms (which determine a given extension of **Z**) be given recursively is non-trivial and cannot be omitted; indeed otherwise, as is noted by Shoenfield[86] (p. 132), we could achieve a complete extension of **Z** simply by adopting as our axioms all statements true in **Z**.

Some Specific Axioms

Let us now consider briefly a few of the axioms adjoined at various times in recent years to the axioms of Zermelo-Fraenkel set theory by topologists (and others), together with some of the reasons these axioms have been chosen. Following standard practice, I will denote by ZF the axioms of Zermelo-Fraenkel set theory and by ZFC the system ZF together with the Axiom of Choice.

§4. The Axiom of Choice

For most of us, this well-known axiom has great intuitive appeal: It is equivalent to the statement that the product of a set of nonempty sets is nonempty and to the statement that a product of compact topological spaces is compact (Kelley[53]). (In contrast, Halpern[44] has shown that the Axiom of Choice does not follow from the statement that the product of compact Hausdorff spaces is compact.) Among the consequences of the Axiom of Choice useful to set theorists is the fact that every filter on a set extends to an ultrafilter; this observation, indeed the first mention of ultrafilters, was made in 1908 by F. Riesz[78] in an address that unfortunately did not receive, at the time, the attention it deserved. To the point-set topologist, the Axiom of Choice is virtually indispensable because it allows (among other things) the definition or construction of the Stone-Čech compactification βX of each completely regular Hausdorff space X. As you are undoubtedly aware, βX is determined, among all compact Hausdorff spaces in which X is densely embedded, as the space to which every continuous function from X to [0,1] (equivalently: to a compact Hausdorff space) extends continuously. We note that, although Čech[13] and Stone[94] have given detailed definitions of the spaces βX in 1937, several aspects of the construction and its properties had been anticipated in 1930 by Tychonoff[99]; see in this connection the remarks of Čech[13] (p. 823) and Stone[95] (p. 26).

The two "applications" of the Axiom of Choice mentioned above—the existence of ultrafilters and the definition of the Stone-Čech compactification—are in fact closely related. In the case of a discrete space X, for example, βX (as a set) may be conveniently identified with the set of all ultrafilters on X (for a systematic treatment from this point of view, see e.g., Gillman and Jerison[39] and Comfort and Negrepontis[22]).

I have attempted in a paper[17] to associate with every completely regular Hausdorff space X, without appealing to the Axiom of Choice, a space with several of the properties of βX; this paper contains references to similar studies by other workers in various areas of mathematics.

Despite the fact that the Axiom of Choice is "intuitively obvious" to many people and despite its many plausible and useful consequences, it must be conceded that certain of its consequences are unpleasant and counterintuitive. For example, using it, one can construct subsets of the real line which are not Lebesgue measurable[101]; and there is the annoying Banach-Tarski paradox[1] (see Stromberg[96] for an elegant and elementary, self-contained treatment) according to which a sphere of radius r may be decomposed into a finite number of pieces (Robinson[79] has shown the minimal number possible is 5) and reassembled into a sphere of radius $2r$. With such anomalies as these, it is not surprising that various mathematicians have looked for alternatives to the Axiom of Choice (for an extensive list of equivalents of the Axiom of Choice, see Rubin and Rubin[80]).

§5. The Axiom of Determinacy

This axiom was not concocted explicitly for the purpose of negating or contradicting the Axiom of Choice, but, as we shall see in a moment, it is, in fact, incompatible with it. To state the Axiom of Determinacy, consider first the following game. Given two players I and II and given $A \subset \omega^\omega$, let I and II define an element f of ω^ω by taking turns defining its entries; that is, I chooses $f(1)$, then II chooses $f(2)$, then I chooses $f(3)$, and so forth. The game is won by I if the sequence f, an element of ω^ω, satisfies $f \in A$; otherwise $f \in \omega^\omega \backslash A$ and the game is won by II. The game just described is denoted $G(A)$. It is said to be *determined* if either I or II has a winning strategy. (Of course, the term *strategy* is a technical expression which requires definition. For present purposes, speaking briefly and informally, let us say simply that a strategy for I is an effective rule defining $f(2n+1)$ if the integers $f(k)$ have been defined for all k such that $0 < k \leq 2n$; a strategy for II is defined analogously.) Now finally we can state the Axiom of Determinacy. It says simply that $G(A)$ is determined for each $A \subset \omega^\omega$.

"That's all well and good," you say, "but isn't an axiom supposed to be 'intuitively obvious,' or at least to reflect some aspect of the real world as we know it? I can see that certain of the games $G(A)$ are determined, for example if A is the open set

$$A = \{ f \in \omega^\omega : f(2) = f(4) = 19 \}$$

then certainly II has a winning strategy, but I do not feel at all that each of the games $G(A)$ is, or in some sense should be, determined." I am neither qualified nor disposed to respond adequately to that criticism, though I imagine that some of those who have worked extensively and productively with the axiom could do so effectively and with pleasure[73,74,75] (see in this connection the spirited, opinionated Introduction to Kleinberg[56]).

In any event it must be admitted that this axiom has some pleasant consequences. With it, for example, every subset of **R** is Lebesgue-measurable,[74] and every uncountable subset of **R** contains a perfect set. Further, I remind you of the important result proved recently by D. A. Martin[71] in ZFC: Every game $G(A)$ with A a Borel set is determined. Thus it may be said that, at the very worst, the Axiom of Determinacy is a pleasing and natural generalization of a theorem provable in ZFC.

The most striking effect of the Axiom of Determinacy, incidentally, is its effect on the behavior of ordinals and cardinals. The cardinals $\aleph_1$ and $\aleph_2$—but not $\aleph_n$, for $2 < n < \omega$—are measurable (in the sense defined in the next section). Combinatorial proofs of these and other consequences of this axiom are given by Kleinberg.[56]

I said earlier that I would not editorialize for or against any particular axiom, but let me go so far as to admit now that I find the Axiom of Determinacy unpleasant and quite unacceptable. My reasons are emotional, self-serving, and inadmissible in a court of law or logic. We know that from this axiom follows the statement that every subset of **R** is Lebesgue-measurable. On the other hand, it is a nice theorem of Sierpiński,[88] proved elegantly and extended by Semadeni,[84] that from a nonprincipal ultrafilter on ω one can define explicitly a nonmeasurable subset of **R**. Thus it follows from the Axiom of Determinacy that there are no nonprincipal ultrafilters on the (discrete) space ω—in symbols: $\beta(\omega)\backslash\omega = \phi$.

Now it happens that I have spent a substantial part of my scholarly career studying and publishing about the space $\beta(\omega)$, with particular emphasis on the intriguing subspace $\beta(\omega)\backslash\omega$. My personal reluctance to accept the Axiom of Determinacy is, then, easily understood: It guarantees that I have spent my time investigating properties of the empty set.

§6. "Large Cardinal" Axioms and Gödel's Axiom $V = L$

We have not taken the time here to state the axioms of Zermelo-Fraenkel set theory, but I do remind you that several of those axioms are devoted to procedures which define new sets from old ones. It is not difficult to see that if, from a model of Zermelo-Fraenkel set theory containing the ordinals, one discards all the sets not given explicitly by the axioms, what remains is a possibly smaller model, indeed the least transitive model, of ZF (denoted L). Gödel's Axiom $V = L$,[42] that every set is constructible (from the axioms), is therefore equiconsistent with ZF. Since both the generalized continuum hypothesis and the Axiom of Choice can be proved from the Axiom $V = L$, they are themselves consistent with ZF. (We note that the Axiom of

Choice is a consequence of the generalized continuum hypothesis. A proof of this result, first announced by Lindenbaum and Tarski,[65] appears in Sierpiński.[89])

The Axiom $V = L$ has been useful in settling a number of questions, in topology (cf. SOUSLIN'S PROBLEM, below) and other branches of mathematics. For some examples in this connection, and for historical/philosophical remarks concerning views of this axiom held by Gödel and some other set theorists, see Devlin.[29]

Although the statement $V = L$ is consistent with ZFC, it is not a theorem of ZFC; indeed it is consistent with ZFC that there are cardinals κ such that $\omega < \kappa < 2^{\omega}$ (see Cohen[14,15,16]).

The statement $V = L$ is also settled in the negative by some (but not all) of the so-called large cardinal axioms. One of these, which has been with us for several decades,[97,100] arose from a question with a measure-theoretic background: Is there a cardinal number α which supports a 2-valued countably additive measure which assigns measure 0 to each singleton and measure 1 to the entire space? (Alternatively: Is there an α such that, for some non-principal ultrafilter p on α we have $\cap\ \mathfrak{F} \in p$ whenever $\mathfrak{F} \subset p$ and $|\mathfrak{F}| \leq_{l} \omega$?) It can be shown that the least such cardinal κ is measurable in the sense that there is a nonprincipal κ-complete ultrafilter on κ, i.e., a nonprincipal ultrafilter p such that $\cap\ \mathfrak{F} \in p$ whenever $\mathfrak{F} \subset p$ and $|\mathfrak{F}| < \kappa$. It is not obvious offhand that an uncountable measurable cardinal would have to be large, but in fact it is very large. Ulam[100] and Tarski[97] showed that any such cardinal α is strongly inaccessible in the sense that α is regular and $2^{\beta} < \alpha$ whenever $\beta < \alpha$, and Hanf and Scott[45] (see also Keisler,[52] Theorem 1) showed that, in fact, such a cardinal α has α strongly inaccessible predecessors.

The statement that there is an uncountable, strongly inaccessible cardinal cannot be proved in ZFC; nor can its consistency with ZFC be proved in ZFC. (For a discussion of this fact, which follows from Gödel's second Incompleteness Theorem, see §9.10 of Shoenfield,[86] or Crossley,[24] pp. 56–88.) The existence of such an inaccessible settles neither CH nor $V = L$; the stronger assumption that there is an uncountable measurable cardinal settles $V = L$ in the negative, but it does not decide CH.

Results deriving from the existence of large cardinals take unexpected forms, with some "looking down" (below the cardinal in question) and some "looking up." I shall cite just one instance of each phenomenon. The first seems particularly surprising in view of the fact, indicated above, that the least measurable cardinal is inaccessible from the small cardinals like ω and $2^{\omega} = \mathbf{c} = |\mathbf{R}|$ required for conventional analysis. As to the expression *strongly compact,* which appears in the statement of the second result below, it is perhaps enough for our purposes to note that every strongly compact cardinal is measurable. Specifically, α is strongly compact if every α-complete filter on every set extends to an α-complete ultrafilter.

THEOREM (Solovay[90]). The system [ZFC + "there is a measurable cardinal"] is consistent if and only if the system [ZFC + "Lebesgue-measure has a countably additive extension defined on every set of reals"] is consistent.

THEOREM (Solovay[91]). Let α be a strongly compact cardinal and β a singular strong limit cardinal such that $\beta > \alpha$. Then $\beta^{+} = 2^{\beta}$.

For remarks on the relative power and utility of certain large cardinal axioms see Drake[31]; Comfort and Negrepontis[22] (especially §8) describes the relationships between several classes of large cardinals. The amusing introduction of Kunen[60] (§7) to a transcendental hierarchy of inaccessibility properties is both fanciful and rigorous.

§7. Martin's Axiom

The last special axiom I want to mention has, at least in one of its formulations, a distinctly topological flavor. In fact, it has been called "the topologists' continuum hypothesis." To set the stage for its statement, let me recall for you one version of the Baire category theorem: In a (locally) compact space, every intersection of countably many dense, open subsets is dense. Or, to put it another way: In a (locally) compact space, no open set except ϕ is the union of countably many nowhere dense subsets. Martin's Axiom[92,72,58] reads as follows: In a compact Hausdorff space with CCC (countable chain condition), no union of $< 2^\omega$ nowhere dense sets has interior. (A space is said to have CCC if it admits no uncountable family of pairwise disjoint, nonempty, open subsets.) In view of the Baire category theorem, Martin's Axiom follows immediately from the continuum hypothesis. Thus Martin's Axiom is interesting only when the continuum hypothesis fails, and a number of mathematicians adopt as a statement of Martin's Axiom the statement above together with the denial of the continuum hypothesis.

We shall cite one use in topology of Martin's Axiom in the next section (for authoritative remarks about this axiom and for additional applications, see Rudin[82] and Shoenfield[87]).

Some Undecidable Topological Statements

§8. CCC $\times$ CCC $=$ CCC?

A space is called *separable* if it has a countable, dense subset. Quite clearly the product of finitely many separable spaces is separable, and it is not difficult to show that the product of countably many separable spaces is separable. (One has to be a little careful here. If D_n is dense in X_n for $n < \omega$, then certainly $\Pi_{n<\omega} D_n$ is dense in $\Pi_{n<\omega} X_n$. But from $|D_n| \leq \omega$, one can conclude not $|\Pi_{n<\omega} D_n| \leq \omega$, but only $|\Pi_{n<\omega} D_n| \leq 2^\omega$. Nevertheless it is not difficult to use $\{D_n : n < \omega\}$ to define a countable dense subset of $\Pi_{n<\omega} X_n$.) Much more suprising, however, is the fact that the product of 2^ω separable spaces is separable. This is the case $\alpha = \omega$ of a result often referred to as the Hewitt-Marczewski-Pondiczery theorem: If $\alpha \geq \omega$, $d(X_i) \leq \alpha$ for $i \in I$, $|I| \leq 2^\alpha$ and $X = \Pi_{i\in I} X_i$, then $d(X) \leq \alpha$. (Here d denotes density character: $d(X)$ is the smallest cardinal number which is the cardinality of a dense subset of X.) A proof of the Hewitt-Marczewski-Pondiczery theorem, together with an appropriately formulated converse and references to the literature concerning density character of product spaces, appears in Comfort and Negrepontis[22] (pp. 77 ff.). My own proof[18] of the case $\alpha = \omega$ seems so simple that we can give an outline here.

Theorem 8.1 Let $|I| \leq 2^\omega$, $d(X_i) \leq \omega$ for $i \in I$ and $X = \Pi_{i\in I} X_i$. Then $d(X) \leq \omega$.

Proof: Since the (discrete) space $\mathbf{Z}$ of integers maps (continuously) onto a dense subspace of X_i, the space $\mathbf{Z}^{(2^\omega)}$ maps continuously onto a dense subspace of X. Hence it is enough to show that $d(\mathbf{Z}^{(2^\omega)}) \leq \omega$. We note first, denoting by $\mathbf{R}$ the real line in its usual topology, that $\mathbf{R}^{(2^\omega)}$, which is naturally identified with $\mathbf{R}^\mathbf{R}$, is separable. Indeed if $\{\langle x_i, y_i\rangle : i \leq n\} \subset \mathbf{R} \times \mathbf{R}$ with $n < \omega$ and with $x_i \neq x_j$ whenever $i \neq j$, and if $\epsilon_i > 0$ for $i \leq n$, then there is a polynomial f with rational coefficients such that

$$| f(x_i) - y_i | < \epsilon_i \text{ for } i \leq n;$$

thus the (countable) set $\mathbf{P}$ of polynomials with rational coefficients is dense in $\mathbf{R}^\mathbf{R}$.

Now with $f \in \mathbf{P}$ associate $\bar{f} : \mathbf{R} \to \mathbf{Z}$ such that

$$| f(x) - \bar{f}(x) | \leq \tfrac{1}{2} \text{ for all } x \in \mathbf{R}.$$

It is clear that $\{ \bar{f} : f \in \mathbf{P}\}$ is dense in $\mathbf{Z}^\mathbf{R}$, so that

$$d(\mathbf{Z}^{(2^\omega)}) \leq |\mathbf{P}| = \omega,$$

as required.

I have dwelt for some moments on the subject of separable spaces because it is clear that every separable space has CCC. Further it has been shown by Marczewski[69] that the product of any set of separable spaces has CCC. (This result is susceptible to substantial generalization. See e.g., Noble and Ulmer[75] or Comfort and Hager[20] Theorem 3.2 or Comfort and Negrepontis[22] Theorem 3.9.) My former students[75] showed that if $\Pi_{i \in F} X_i$ has CCC for every finite $F \subset I$, then in fact $\Pi_{i \in I} X_i$ has CCC (again there are substantial generalizations available; see e.g., Comfort and Negrepontis,[22] §3).

These results bring to mind the following natural question: Is the product of finitely many spaces with CCC a space with CCC? Symbolically: CCC $\times$ CCC $=$ CCC? This sounds like a natural question not excessively difficult to solve—it was given to the topology class I attended in graduate school, and was characterized then as "challenging"—but in fact it is undecidable in ZFC. We can understand the situation by considering this question together with another, even better-known, question.

§9. Souslin's Problem

It is a well-known theorem, dating back to Cantor,[12] that the real line $\mathbf{R}$ (i) is densely ordered, (ii) is order-complete, (iii) has no first or last element, and (iv) is separable; further, $\mathbf{R}$ is, up to order-isomorphism, the only such ordered space. (Note that this is an "order-theoretic" theorem, or a topological theorem if you prefer, not an algebraic theorem; we are ignoring the algebraic properties of $\mathbf{R}$ for the moment.) The question raised by Souslin[93] is this: can (iv) be replaced by (iv′) has CCC? Note that Souslin was not asking whether $\mathbf{R}$ has CCC—surely it does—nor whether CCC is, in general, equivalent to separable (since he knew it was not). Rather, Souslin was asking whether, in the presence of the other three conditions, separability and CCC are equivalent conditions. This is *Souslin's problem;* a *Souslin space* is a space which responds negatively to Souslin's problem—i.e., a nonseparable space with properties (i), (ii), (iii), and (iv′).

What is the relation between Souslin's problem and the question whether the product of two spaces with CCC has CCC? Kurepa[61,62] showed that if S is a *Souslin line* then (S has CCC and) $S \times S$ does not have CCC. More recently, Tennenbaum[98] and Jech[48] have proved the consistency with ZFC of the existence of a Souslin line; in fact, Tennenbaum proved the consistency even with ZFC + GCH or, alternatively, with ZFC $+\neg$CH. In an important, seminal proof which gave rise to the formulation of the combinatorial principle ◊ (diamond) and which brings into focus the connection between the recursive definition of Gödel's constructible universe L and pure combinatorics, Jensen[50] derived the existence of a Souslin line from the axiom $V = L$. (The details of Jensen's work are exposed in Devlin[30].)

In an unpublished work, R. Laver showed that if $\omega^+ = 2^\omega$, then there is a space X with CCC such that $X \times X$ does not have CCC. Galvin[38] has given a simple proof of Laver's result, together with a number of generalizations.

In contrast with these results, it has been shown by Solovay and Rowbottom, and independently by Kunen, that under Martin's Axiom and the denial of the continuum hypothesis every space with CCC has this property: Among every ω^+ nonempty open subsets, some ω^+ have the finite intersection property. From this it follows immediately (assuming Martin's Axiom and the denial of the continuum hypotheses) that the product of finitely many spaces with CCC has CCC (proofs of these results are in Juhász[51]). In another major result, whose proof occupies the latter half of Devlin and Johnsbråten,[30] Jensen shows the consistency with ZFC + CH of the statement that there is no Souslin space.

The preceding few paragraphs indicate, among other things, that two famous problems in general topology are undecidable on the basis of the axioms of ZFC. Specifically, there are models of ZFC in which there is no Souslin line and in which every product of CCC spaces has CCC, and there are models of ZFC in which there is a Souslin line (which has CCC and whose square does not). In this respect the problems in question are similar to those involving Euclid's fifth postulate, described earlier, and perhaps to Fermat's Last Theorem.

It is well to note, however, that there are differences: In our everyday model of Euclidean 3-space, with "point," "line," and "parallel" assigned their usual interpretations, Euclid's fifth postulate is emphatically true and easily proved, and Fermat's Last Theorem, if indeed it is undecidable, is true in our model of the integers. In contrast, there is at this moment no special or privileged model of ZFC (commonly agreed upon by mathematicians) in which Souslin's problem or the CCC question may be said to be settled. We have seen that there are models, not necessarily bizarre or counterintuitive, in which these questions are answered positively, and others in which the answer is negative. For more complete and authoritative accounts of Souslin's problem and its descendants, see Rudin[81,82] and Devlin and Johnsbråten.[30] This problem, together with the CCC $\times$ CCC = CCC question and other axiom-sensitive topological statements, is examined carefully and at length from an elementary point of view by Bennett and McLaughlin.[2]

§10. The Frechet Order on ω^ω

It occurs frequently that well-posed, unambiguous questions about well-defined mathematical objects cannot be answered absolutely. A striking case in point arises in connection with the set ω^ω of functions from ω to ω with the Fréchet order $<$ defined by

$$f < g \text{ if } |\{n < \omega : g(n) \leq f(n)\}| < \omega.$$

It is easily checked that $<$ is a partial order on the set ω^ω.

Among the questions not settled by the axioms of ZFC are these two.

(1) Is there a cofinal, linearly ordered subset of ω^ω?
(2) Is there a one-to-one, order-preserving function from ω^ω onto a linearly ordered subset of ω^ω?

Question (2) has been of particular interest because of its role in the recent solution by Solovay and Woodin of Kaplansky's question: Is there a discontinuous homomorphism from the Banach algebra $C([0, 1])$ of complex-valued continuous functions defined on the closed interval $[0, 1]$ into some Banach algebra? (This question, answered affirmatively by Dales[25] and Esterle[33] assuming the continuum hypothesis, arose naturally out of the classical theorem of Gelfand to the effect that every complex-valued homomorphism on $C([0, 1])$ is indeed continuous.) Woodin showed that in any model of ZFC with a discontinuous homomorphism of this kind, question (2) has a positive answer; and Solovay, using Martin's Axiom and the forcing method of Cohen, defined a model of ZFC in which question (2) has a negative answer.

I am grateful to Professor T. Jech for sending me a copy of his classroom lecture-notes[49] on the work of Solovay and Woodin and the discontinuous homomorphism problem.

With the partially ordered set $\langle\omega^\omega, <\rangle$ before us I cannot resist taking a moment to mention my favorite theorem concerning it. To do this, let us first consider the following four cardinal numbers: (i) the minimal cardinality of a cofinal subset of ω^ω; (ii) the minimal number of compact subspaces of **R** whose union is the set of irrational numbers; (iii) the minimal cardinal κ such that the space of rational numbers embeds as a closed subspace of ω^κ; (iv) the minimal number of open-and-closed sets required to cover the interior of a nonopen zero-set in $\beta(\omega)\backslash\omega$. It is not difficult to see that each of these cardinal numbers is uncountable and $\leq 2^\omega$. The theorem I admire, due in most of its essential features to Hechler,[46] is that they are all equal. Further, according to Hechler, it is consistent with ZFC that this cardinal assume any value λ such that $cf(\lambda) > \omega$ and $\lambda \leq 2^\omega$.

The equality between (i) and (ii) above is used by Comfort and Negrepontis[23] (Corollary 11.8) to show that not every Baire set in a paracompact space is paracompact and, assuming the continuum hypothesis, that not every Baire set in a Lindelöf space is Lindelöf.

In a survey article[19] I have presented an argument of B. Balcar, R. Frankiewicz, and P. Simon, based on properties of the partially ordered set $\langle\omega^\omega, <\rangle$, showing that if α and γ are cardinals such that $\alpha > \gamma \geq \omega$ and the Stone-Čech remainders $\beta(\alpha)\backslash\alpha$

and $\beta(\gamma)\backslash\gamma$ are homeomorphic, then $\gamma = \omega$ and $\alpha = \omega^+$. (Additional information and references about this simply defined set are given by Rudin,[82] §VII).

§11. The Partially Ordered Set $\beta(\alpha)$

There is another partially ordered set, familiar to topologists acquainted with the theory of ultrafilters, whose order-theoretic properties are not completely determined by the axioms of ZFC. For an infinite cardinal α we denote, as usual, by $\beta(\alpha)$ the Stone-Čech compactification of the (discrete) space α. For $f \in \alpha^\alpha$ we denote by $\overline{f}$ the (unique) continuous function from $\beta(\alpha)$ into $\beta(\alpha)$ such that $f \subset \overline{f}$. The *Rudin-Keisler order* $\preccurlyeq$ is defined on $\beta(\alpha)$ as follows: $p \preccurlyeq q$ if there is $f \in \alpha^\alpha$ such that $\overline{f}(q) = p$. Strictly speaking, the relation $\preccurlyeq$ is not a true partial order on $\beta(\alpha)$, since, from $p \preccurlyeq q$ and $q \preccurlyeq p$, it does not follow that $p = q$. There is a natural equivalence relation $\sim$ on $\beta(\alpha)$ which $\preccurlyeq$ respects, and it is $\langle \beta(\alpha)/\sim, \preccurlyeq \rangle$ which is a partially ordered set. These and other technicalities need not concern us now (a careful, rigorous treatment of the relation $\preccurlyeq$ and its properties is available in Comfort and Negrepontis,[22] §9).

Many aspects of the space $\beta(\alpha)$ are well understood. We know, for example, its cardinality, weight, and density character; Shelah[85] has shown recently that there is a set S of pairwise $\preccurlyeq$-incomparable elements of $\beta(\alpha)$ such that $|S| = 2^{2^\alpha}$. The very natural question I want to cite about the ordered set $\langle \beta(\alpha), \preccurlyeq \rangle$, which is similar to those asked above concerning $\langle \omega^\omega, < \rangle$, is this: Is there a cofinal, linearly ordered subset?

Let us see how different (consistent) axioms can be adjoined to ZFC to produce different answers to this question. We first need a theorem of Hajnal[43]; here $\mathcal{P}(\alpha)$ is the power set of α and $\mathcal{P}_\kappa(\alpha)$ denotes $\{A \subset \alpha : |A| < \kappa\}$.

Theorem 11.1. If $f: \alpha \to \mathcal{P}_\kappa(\alpha)$ with $\omega \leq \kappa < \alpha$ and with $\xi \notin f(\xi)$ for $\xi < \alpha$, then there is $A \subset \alpha$ such that $|A| = \alpha$ and $\xi \notin f(\zeta)$ for $\xi, \zeta \in A$.

Corollary 11.2. Let $\omega \leq \alpha$ and $(2^\alpha)^+ < 2^{2^\alpha}$. Then if $S \subset \beta(\alpha)$ and $|S| = 2^{2^\alpha}$, there is $T \subset S$ such that $|T| = 2^{2^\alpha}$ and the elements of T are pairwise $\preccurlyeq$-incomparable.

Proof: For $q \in S$ set $f(q) = \{p \in S : p \preccurlyeq q \text{ and } p \neq q\}$. Since

$$|f(q)| \leq 2^\alpha < (2^\alpha)^+ < 2^{2^\alpha}$$

we have from Hajnal's theorem (replacing κ and α by $(2^\alpha)^+$ and 2^{2^α}, respectively) that there is $T \subset S$ such that $|T| = 2^{2^\alpha}$ and $p \notin f(q)$ for $p, q \in T$. Clearly the set T is as required.

In contrast to this result, Negrepontis and I noted the following facts.[21]

Theorem 11.3. If $\alpha \geq \omega$, $S \subset \beta(\alpha)$, and $|S| \leq 2^\alpha$, then there is $q \in \beta(\alpha)$ such that $p \preccurlyeq q$ for all $p \in S$.

Proof: Let $S = \{p_\xi : \xi < 2^\alpha\}$, with repetitions allowed, and let

$$p = \langle p_\xi : \xi < 2^\alpha \rangle \in \beta(\alpha)^{(2^\alpha)}$$

According to the Hewitt-Marczewski-Pondiczery theorem cited above, there is a (continuous) function f from α onto a dense subspace of $\alpha^{(2^\alpha)}$. Let g denote its continuous extension from $\beta(\alpha)$ onto $\beta(\alpha)^{(2^\alpha)}$, and π_ξ the projection from $\alpha^{(2^\alpha)}$ onto $\alpha_\xi = \alpha$. Then, with q chosen in $\beta(\alpha)$ so that $g(q) = p$, we have $\pi_\xi \circ f \in \alpha^\alpha$ and

$$p_\xi = \bar{\pi}_\xi(p) = \bar{\pi}_\xi(g(q)) = (\pi_\xi \circ f)^-\ (q)$$

and hence $p_\xi \leqslant q$, as required.

Corollary 11.4. If $\alpha \geq \omega$ and $(2^\alpha)^+ = 2^{2^\alpha}$, then there is a $\leqslant$-linearly ordered, cofinal subset of $\beta(\alpha)$ (of cardinality 2^{2^α}).

Proof: Let $\beta(\alpha) = \{p_\xi : \xi < 2^{2^\alpha}\}$, choose $q_0 \in \beta(\alpha)$ so that $p_0 \leqslant q_0$ and recursively, if $0 < \xi < 2^{2^\alpha}$ and q_η has been defined for all $\eta < \xi$, choose q_ξ so that $p_\xi \leqslant q_\xi$ and $q_\eta \leqslant q_\xi$, for all $\eta < \xi$. Clearly $\{q_\xi : \xi < 2^{2^\alpha}\}$ is as required.

We note that if $(2^\alpha)^+ = 2^{2^\alpha}$ then no subset of $\beta(\alpha)$ of cardinality less than 2^{2^α} can be $\leqslant$-cofinal; for $|\beta(\alpha)| = 2^{2^\alpha}$, and

$$|\{p \in \beta(\alpha) : p \leqslant q\}| \leq 2^\alpha, \text{ for all } q \in \beta(\alpha).$$

From Corollaries 11.2 and 11.4, we see that the two hypotheses $(2^\alpha)^+ < 2^{2^\alpha}$ and $(2^\alpha)^+ = 2^{2^\alpha}$, each of which is consistent with ZFC,[32] produce remarkably different results about the $\leqslant$-structure of $\beta(\alpha)$. From the first it follows that every subset of $\beta(\alpha)$ of cardinality 2^{2^α} contains an equally large subset of $\leqslant$-incomparable elements, while from the second it follows that there is a linearly ordered (cofinal) subset of cardinality 2^{2^α} (which has, of course, no pair of incomparable elements).

This used to disturb me a great deal. In contrast with that CCC question, which one can accept quite readily as being determined in different ways by different axiomatic assumptions, I felt that the Stone-Čech compactification of a discrete space α does really exist and is concrete. Simple questions about its order-theoretic structure should be answerable absolutely. In fact, however, the differing answers to the same question which are afforded by Corollaries 11.2 and 11.4 become less disturbing when it is noted that the question itself can be recast in language which concerns only the cardinal numbers α, 2^α, 2^{2^α}, and some of their subsets. It is hardly surprising that such a question may have two different answers depending on whether $(2^\alpha)^+$ is or is not equal to 2^{2^α}.

§12. Weiss's Solution to the Blumberg Problem

Before closing I want to cite a question whose solution turned out not to be axiom-sensitive. To set the stage, I recall first a remarkable old theorem of Blumberg[4]: If f is any function from **R** to **R**, then there is dense $D \subset \mathbf{R}$ such that $f|D$ is continuous. (This theorem is occasionally misunderstood. One might be tempted to propose as a counterexample the characteristic function of the rationals; that function is not continuous at any point of **R**. But it is not a counterexample to Blumberg's theorem since the set of rationals is a dense subset of **R** on which it is constant and hence continuous.) Let us say, in keeping with terminology now widely accepted, that a space

X is a *Blumberg space* if for every $f: X \to \mathbf{R}$, there is dense $D \subset X$ such that $f | D$ is continuous. Blumberg's original paper[4] showed that, in fact, every complete, separable metric space is a Blumberg space. Later Bradford and Goffman[6] achieved the same conclusion for Baire spaces, and the question became current in the early 1970's whether every compact Hausdorff space is a Blumberg space.

Using some techniques introduced by White[103] and Levy,[63,64] as well as a number of fresh ideas of his own, Weiss[102] answered this question in an interesting way. Working in ZFC, he defined two compact Hausdorff spaces (let us call them X and Y) with these properties: If CH, then X is not a Blumberg space, and if $\neg$ CH, then Y is not a Blumberg space. It is then clear that $X \times Y$, or, alternatively, the disjoint union $X \uplus Y$, is a compact Hausdorff space which is not a Blumberg space.

For an understanding of the validity of Weiss's argument, it is important to recognize that neither X nor Y depend for their definition on CH or $\neg$ CH. Both X and Y are well defined in ZFC (indeed X is the Stone space of a certain familiar Boolean algebra, and Y is a Dedekind-complete linearly ordered topological space). Thus in any model of ZFC the space $X \uplus Y$ is a well-defined compact Hausdorff space. In that model either CH holds or CH fails, and accordingly either X or Y is not a Blumberg space; hence $X \uplus Y$ is not a Blumberg space.

§13. Closing Comments

Mathematics shares with all of Science the duty and the goal to examine and describe the real world and our greater environment. This is, presumably, an open-ended task of potentially infinite duration; we shall never know all that might be known. Scientists approach the problem with the tools appropriate to their personal disciplines. My own view, somewhat pessimistic and tinged with embarrassment, is that we have contributed less to the success of this enterprise than, say, the physicists or the biologists. True, we are able to formulate axioms or principles that we hope will shed light on the problem. But when they fail to do so, we revert to the playing of games, to establishing statements which, though true in the formal sense that they follow logically from our assumptions, have no known relationship or relevance to our lives. We have been unable to determine the truth or falsity in the real world of a number of natural, simple statements (e.g., the continuum hypothesis); we have been unable to define with clarity a mathematical model closely approximating the real world.

It is easy for the mathematician to abandon scientific inquiry into the nature of the real world and to retreat to the comfortable position that there is and can be no truth outside of mathematics and that competing mathematical truths, so long as they are derived from one or another set of axioms by legitimate logical processes, are of equal interest and worth. To adopt this stance is to cheapen mathematics and to contribute to the legitimacy of Bertrand Russell's[83] aphorism that "... mathematics may be defined as the subject in which we never know what we are talking about, nor whether what we are saying is true."

To the mathematician who believes that mathematics is quite properly divorced from the real world I ask questions like these: If it were learned tomorrow that the continuum hypothesis holds in the real world, would the theorems based on Martin's Axiom and the denial of the continuum hypothesis retain their interest? If it develops that there are no uncountable measurable cardinals, will the beautiful characterizations (due to Rowbottom, Kunen, and others) of Ramsey ultrafilters, which exist only over measurable cardinals, continue to excite our enthusiasm? The answer to the first of these questions, and I think to the second also, is "No."

We need not despair. Recent developments in logic and set theory of the kind we have touched on here have generated an explosive excitement in general topology (a field written off as dead or dying a decade ago by many mathematicians), and I look forward with confidence to the continued publication of new truths in set theory and topology which are sufficiently deep to be interesting and sufficiently simple and elegant that I can understand and appreciate them.

Acknowledgments

I am grateful to several mathematicians for reading, carefully and critically, an early version of this manuscript. In this connection I am indebted most particularly to Garrett Birkhoff, Robert Rosenbaum, Stanley Wagon, and the referees. Each of these uncovered a number of errors and misperceptions on my part, some petty and some major; and each made some imaginative suggestions designed to enhance the scope and the reliability of this paper.

References

1. Banach, S. & A. Tarski. 1924. Sur la décomposition des ensembles de points en parties respectivement congruentes. Fundam. Math. **6:** 244–277.
2. Bennett, H. R. & T. G. McLaughlin. 1976. A Selective Survey of Axiom-Sensitive Results in General Topology. Texas Tech University Mathematics Series Vol. 12. Texas Tech University, Lubbock, Texas.
3. Bishop, E. 1967. Foundations of Constructive Analysis. McGraw-Hill Book Co., New York, N.Y.
4. Blumberg, H. 1922. New properties of all real functions. Trans. Am. Math. Soc. **24:** 113–128.
5. Bolyai, J. 1832. Scientiam spatii absolute veran exhibens: a veritate aut falsitate Axiomatis XI Euclidei independentem. Maros-Vásárhelyini, Hungary.
6. Bradford, J. C. & C. Goffman. 1960. Metric spaces in which Blumberg's theorem holds. Proc Am. Math. Soc. **11:** 667–670.
7. Brouwer, L. E. J. 1907. Over de Grondslagen der Wiskunde (On the foundations of mathematics). Amsterdam, the Netherlands. Doctoral dissertation.
8. Brouwer, L. E. J. 1908. Over de Onbetrouwbaarheid der logische Principes (On the untrustworthiness of logical principles). Tijdscr. Wijsbe. **2:** 152–158.
9. Brouwer, L. E. J. 1974. Collected works. Vol. 1. Philosophy and Foundations of Mathematics. A. Heyting, Ed. North-Holland/American Elsevier, Amsterdam, the Netherlands/New York, N.Y.
10. Brouwer, L. E. J. 1976. Collected works. Vol. 2. Geometry, Analysis, Topology and Mechanics. H. Freudenthal, Ed. North-Holland/American Elsevier, Amsterdam, the Netherlands/New York, N.Y.
11. Cantor, G. 1892. Uber eine elementare Frage der Mannigfaltigkeitslehre. Jahresber. Dtsch. Mathematikerver. **1:** 75–78
12. Cantor, G. 1895. Beiträge zur Begründund der transfinite Mengen. Math. Ann. **46:** 481–512.
13. Čech, E. 1937. On bicompact spaces. Ann. Math. (2) **38:** 823–844.
14. Cohen, P. J. 1963. The independence of the continuum hypothesis I. Proc. Nat. Acad. Sci. U.S.A. **50:** 1143–1148.
15. Cohen, P. J. 1964. The independence of the continuum hypothesis, II. Proc. Nat. Acad. Sci. U.S.A. **51:** 105–110.

16. COHEN, P. J. 1966. Set Theory and the Continuum Hypothesis. W. A. Benjamin, Inc, New York, N.Y./Amsterdam, the Netherlands.
17. COMFORT, W. 1968. A Theorem of Stone-Čech type, and a theorem of Tychonoff type, without the axiom of choice; and their realcompact analogues. Fundam. Math. **63:** 97–110.
18. COMFORT, W. W. 1969. A short proof of Marczewski's separability theorem. Am. Math. Mon. **9:** 1041–1042.
19. COMFORT, W. W. 1977. Compatifications: recent results from several countries. Proc. 1977 South. Topology Conf. La. State Univ. Topology Proc. **2:** 61–87.
20. COMFORT, W. W. & A. W. HAGER. 1970. Estimates for the number of real-valued continuous functions. Trans. Am. Math. Soc. **150:** 619–631.
21. COMFORT, W. W. & S. NEGREPONTIS. 1972. On families of large oscillation. Fundam. Math. **75:** 275–290.
22. COMFORT, W. W. & S. NEGREPONTIS. 1974. The Theory of Ultrafilters. Grundlehren math. Wiss. Vol. 211. Springer-Verlag. Berlin-Heidelberg, Federal Republic of Germany/New York, N.Y.
23. COMFORT, W. W. & S. NEGREPONTIS. 1975. Continuous Pseudometrics. Lect. Notes Pure Appl. Math. Vol. 14. Marcel Dekker, New York, N.Y.
24. CROSSLEY, J. N. *et al.* 1972. What is Mathematical Logic? Oxford University Press, London-Oxford, England/New York, N.Y.
25. DALES, H. G. 1979. Manuscript. Am. J. Math. To appear.
26. DAVIS, M. 1973. Hilbert's tenth problem is unsolvable. Am. Math. Mon. **80:** 233–269.
27. DAVIS, M. 1977. Unsolvable problems. *In* Handbook of Mathematical Logic. North-Holland, Amsterdam, the Netherlands/New York, N.Y. pp. 567–594.
28. DAVIS, M., YU MATIJASEVIČ & J. ROBINSON. 1976. Hilbert's tenth problem. Diophantine equations: positive aspects of a negative solution. *In* Mathematical Developments Arising from Hilbert Problems. Proc. Symp. Pure Math. American Mathematical Society. **28:** 323–378. Providence, R.I.
29. DEVLIN, K. J. 1977. The axiom of constructibility: A guide for the mathematician. Lect. Notes Math. Vol. 617. Springer-Verlag, Berlin-Heidelberg, Federal Republic of Germany/New York, N.Y.
30. DEVLIN, K. J. & H. JOHNSBRÅTEN. 1974. The Souslin problem. Lect. Notes Math. Vol. 405. Springer-Verlag, Berlin-Heidelberg, Federal Republic of Germany/New York, N.Y.
31. DRAKE, F. R. 1974. Set theory an introduction to large cardinals. Studies in Logic and the Foundations of Mathematics Vol. 76. North-Holland/American Elsevier, Amsterdam, the Netherlands/London, England/New York, N.Y.
32. EASTON, W. B. 1970. Powers of regular cardinals. Ann. Math. Logic **1:** 139–178.
33. ESTERLE, J. 1979. Manuscript. Proc. London Math. Soc. To appear.
34. EUCLID OF ALEXANDRIA. 1883 edit. Euclidis opera omnia. Vol. 1–5. J. L. Heiberg & H. Menge, Eds. Teubner Classical Library, Leipzig, Federal Republic of Germany.
35. FRAENKEL, A. A. 1922. Zu den Grundlagen der Cantor-Zermaloschen Mengenlehre. Math. Ann. **86:** 230–237.
36. FRAENKEL, A. A. 1922. Der Begriff "definit" und die Unabhängigkeit des Answahlaxioms. Sitzungsber. Preuss. Akad. Wiss. Phys.-math. Kl. pp. 253–257.
37. FREUDENTHAL, H. & A. HEYTING. 1976. The life of L. E. J. Brouwer. *In* L. E. J. Brouwer, Collected Works. Vol. 2. H. Freudenthal, Ed. North-Holland/American Elsevier, Amsterdam, the Netherlands/New York, N.Y. pp. x–xv.
38. GALVIN, F. 1979. Chain conditions and products. Fundam. Math. To appear.
39. GILLMAN, L. & M. JERISON. 1960. Rings of Continuous Functions. D. Van Nostrand, Princeton, N.J.
40. GÖDEL, K. 1931. Über formal unentscheidbare Sätze der Principia Mathematicae und verwandter Systeme, I. Monatsh. Math. Phys. **38:** 173–198.
41. GÖDEL, K. 1938. The consistency of the axiom of choice and the generalized continuum hypothesis. Proc. Nat. Acad. Sci. U.S.A. **24:** 556–557.
42. GÖDEL, K. 1940. The Consistency of the Axiom of Choice and of the Generalized Continuum Hypothesis with Axioms of Set Theory. Ann. Math. Stud. No. 3. Princeton University Press, Princeton, N.J.
43. HAJNAL, A. 1961. Proof of a conjecture of S. Ruziewicz. Fundam. Math. **50:** 123–128.

44. HALPERN, J. D. 1964. The independence of the axiom of choice from the Boolean prime ideal theorem. Fundam. Math. **55:** 57–66.
45. HANF, W. P. & D. SCOTT. 1961. Classifying inaccessible cardinals. Not. Am. Math. Soc. **8:** 445.
46. HECHLER, S. H. 1975. On a ubiquitous cardinal. Proc. Am. Math. Soc. **52:** 348–352.
47. HILBERT, D. 1900. Mathematische Probleme, Vortrag, gehalten auf dem internationalen Mathematiker-Kongress zu Paris 1900. Nachr. Akad. Wiss. Göttingen, Math.-Phys. Kl. pp. 253–297. English translation: Bull. Am. Math. (1901): **8:** 437–479.
48. JECH, T. J. 1967. Non-provability of Souslin's hypothesis. Comment. Math. Univ. Carol. **8:** 291–305.
49. JECH, T. 1977. The discontinuous homomorphism problem. Classroom notes.
50. JENSEN, R. B. 1968. Souslin's hypothesis is incompatible with $V = L$. (Abstract). Not. Am. Math. Soc. **15:** 935.
51. JUHÁSZ, I. 1970. Martin's axiom solves Ponomarev's problem. Bull. Acad. Polon. Sci. Sér. Sci. Math. Astron. Phys. **18:** 71–74.
52. KEISLER, H. J. 1962. Some applications of the theory of models to set theory. *In* Logic, Methodology and Philosophy of Science. Proc. 1960 Int. Congr. E. Nagel, Ed. Stanford University Press, Stanford, Cal. pp. 80–86.
53. KELLEY, J. L. 1950. The Tychonoff theorem implies the axiom of choice. Fundam. Math. **37:** 75–76.
54. KLEIN, F. 1871. Über die sogenannte Nicht-Euklidische Geometrie. Math. Ann. **4:** 573–625.
55. KLEIN, F. 1873. Über die sogenannte Nicht-Euklidische Geometrie, II. Math. Ann. **6:** 112–145.
56. KLEINBERG, E. M. 1977. Infinitary combinatorics and the axiom of determinateness. Lect. Not. Math. Vol. 612. Springer-Verlag, Berlin-Heidelberg, Federal Republic of Germany/New York, N.Y.
57. KREISEL, G. 1977. Review of L. E. J. Brouwer collected works. Vol. 1. Bull. Am. Math. Soc. **83:** 86–93.
58. KUNEN, K. 1968. Inaccessibility properties of cardinals. Doctoral dissertation. Stanford University, Stanford, Cal.
59. KUNEN, K. 1972. Ultrafilters and independent sets. Trans. Am. Math. Soc. **172:** 299–306.
60. KUNEN, K. 1978. Combinatorics. *In* Handbook of Mathematical Logic. North-Holland, Amsterdam, the Netherlands/New York, N.Y. pp. 371–400.
61. KUREPA, G. 1950. La condition de Souslin et une propriété caractéristique des nombres réels. C. R. Acad. Sci. Paris **231:** 1113–1114.
62. KUREPA, G. 1952. Sur une propriété caractéristique du continu linéaire et le problème de Suslin. Acad. Sci. Serbe Publ. Inst. Math. **4:** 97–108.
63. LEVY, R. 1973. A totally ordered Baire space for which Blumberg's theorem fails. Proc. Am. Math. Soc. **41:** 304.
64. LEVY, R. 1974. Strongly non-Blumberg spaces. Gen. Topology Appl. **4:** 173–177.
65. LINDENBAUM, A. & A. TARSKI. 1926. Communication sur les recherches de la théorie des ensembles. C. R. Varsovie **19:** 299–330.
66. LOBACHEVSKY, N. I. 1836. Neue Anfangsgruende de Geometrie, mit einer vollständigen Theorie der Parallelen. Gelehrte Schriften der Universität Kasan, Kasan, U.S.S.R.
67. LOBACHEVSKY, N. I. 1837. Geometrie imaginaire. Crelle's J. **17:** 295–320.
68. ŁOŚ, J. & C. RYLL-NARDZEWSKI. 1955. Effectiveness of the representation theory for Boolean algebras. Fundam. Math. **41:** 49–56.
69. MARCZEWSKI, E. 1947. Separabilité et multiplication cartésienne des espaces topologiques. Fundam. Math. **34:** 127–143.
70. MARTIN, D. A. 1968. The axiom of determinateness and reduction principles in the analytical hierarchy. Bull. Am. Math. Soc. **74:** 687–689.
71. MARTIN, D. A. 1975. Borel determinacy. Ann. Math. **102:** 363–371.
72. MARTIN, D. A. & R. M. SOLOVAY. 1970. Internal Cohen extensions. Ann. Math. Logic **2:** 143–178.
73. MYCIELSKI, J. 1964. On the axiom of determinateness. Fundam. Math. **53:** 205–224.
74. MYCIELSKI, J. & S. SWIERCZKOWSKI. 1964. On the Lebesgue measurability and the axiom of determinateness. Fundam. Math. **54:** 67–71.

75. Noble, N. & M. Ulmer. 1972. Factoring functions on Cartesian products. Trans. Am. Math. Soc. **163:** 329–340.
76. Paris, J. & L. Harrington. 1978. A mathematical incompleteness in Peano arithmetic. *In* Handbook of Mathematical Logic. North-Holland, Amsterdam, the Netherlands/New York, N.Y. pp. 1133–1142.
77. Peano, G. 1889. Arithmetices Principia, Nova Methodo Exposita. Bocca, Turino, Italy.
78. Riesz, F. 1909. Stetigkeitsbegriff und abstrakte Mengenlehre. *In* Atti IV Cong. Int. Mat., Rome 1908. G. Castelnuova, Ed. Tipografia R. Accad. Lincei. Rome, Italy. **11:** 18–24.
79. Robinson, R. M. 1947. On the decomposition of spheres. Fundam. Math. **34:** 246–260.
80. Rubin, H. & J. Rubin. 1963. Equivalents of the Axiom of Choice. North-Holland, Amsterdam, the Netherlands.
81. Rudin, M. E. 1969. Souslin's conjecture. Am. Math. Mon. **76:** 113–119.
82. Rudin, M. E. 1975. Lectures on set theoretic topology. Reg. Conf. Ser. Math. Vol. 23. American Mathematical Society, Providence, R. I.
83. Russell, B. 1901. Recent works on the principles of mathematics. Int. Mon. **4:** 83–101.
84. Semadeni, A. 1964. Periods of measurable functions and the Stone-Čech compactification. Am. Math. Mon. **71:** 891–893.
85. Shelah, S. 1978. Unpublished manuscript. To appear.
86. Shoenfield, J. R. 1967. Mathematical Logic. Addison-Wesley, Reading, Mass.
87. Shoenfield, J. R. 1975. Martin's axiom. Am. Math. Mon. **82:** 610–617.
88. Sierpiński, W. 1938. Fonctions additives non complètement additives et fonctions non mesurables. Fundam. Math. **30:** 96–99.
89. Sierpiński, W. 1947. L'hypothèse géneralisée du continu et l'axiome du choix. Fundam. Math. **34:** 1–5.
90. Solovay, R. M. 1971. Real-valued measurable cardinals. Axiomatic Set Theory. Proc. Symp. Pure Math. part 1. D. S. Scott, Ed. American Mathematical Society, Providence, R.I. **13:** 397–428.
91. Solovay, R. M. 1974. Strongly compact cardinals and the GCH. Proc. Symp. Pure Math. Proc. Tarski Symp. L. Henkin *et al.,* Eds. American Mathematical Society. **25:** 365–372.
92. Solovay, R. M. & S. Tennenbaum. 1971. Iterated Cohen extensions and Souslin's problem. Ann. Math. **94:** (2) 201–245.
93. Souslin, M. 1920. Problème 3. Fundam. Math. **1:** 223.
94. Stone, M. H. 1937. Applications of the theory of Boolean rings to general topology. Trans. Am. Math. Soc. **41:** 375–481.
95. Stone, M. H. 1962. Commemoration of Eduard Čech. *In* General Topology and Its Relations to Modern Analysis and Algebra. Proc. 1961 Prague Topological Symp. J. Novák, Ed. Czechoslovak Academy of Sciences, Prague, Czechoslovakia. pp. 26–28.
96. Stromberg, K. R. 1979. The Banach-Tarski paradox. To appear.
97. Tarski, A. 1938. Drei Überdeckungssätze der allgemeinen Mengenlehre. Fundam. Math. **30:** 132–155.
98. Tennenbaum, S. 1968. Souslin's problem. Proc. Nat. Acad. Sci. U.S.A. **59:** 60–63.
99. Tychonoff, A. 1930. Über die topologische Erweiterung von Räumen. Math. Ann. **102:** 544–561.
100. Ulam, S. 1930. Zur Masstheorie in der allgemeinen Mengenlehre. Fundam. Math. **16:** 140–150.
101. Vitali, G. 1905. Sul Problema della Misura dei Gruppi di Punti di una Retta. Bologna, Italy.
102. Weiss, W. A. R. 1975. A solution to the Blumberg problem. Bull. Am. Math. Soc. **81:** 957–958.
103. White, H. E., Jr. 1974. Topological spaces in which Blumberg's theorem holds. Proc. Am. Math. Soc. **44:** 454–462.
104. Zermelo, E. 1908. Untersuchengen über die Grundlagen der Mengenlehre, I. Math. Ann. **65:** 261–281.

GEORG CANTOR'S CREATION OF TRANSFINITE SET THEORY: PERSONALITY AND PSYCHOLOGY IN THE HISTORY OF MATHEMATICS*†

Joseph W. Dauben

Department of History
Herbert H. Lehman College
City University of New York
Bronx, New York 10468

Historians of mathematics are generally accustomed to discussing ideas rather than individuals. A mathematician's biography and his mathematics are frequently regarded as wholly separate, the former providing human interest (perhaps), the latter however comprising in most cases the heart of the matter. But the analysis of personality, in particular that of creative individuals, even mathematicians, can reveal a great deal about the nature of discovery and the character of the otherwise disembodied ideas with which they deal.

In the case of the German mathematician, Georg Cantor, this is especially true. If one were to consult only the published record of his research, the factors influencing his discovery and subsequent development of set theory and the transfinite numbers would remain obscure. At best one could hope for only a partial, and probably inaccurate, view of what Cantor accomplished. Why should he have been the mathematician most likely to defend transfinite set theory, despite tremendous opposition not only from mathematicians, but from philosophers and theologians as well? How did his character and scientific temperament shape the earliest development of set theory? Satisfying answers to such questions cannot be expected without going beyond the record of Cantor's printed papers. In letters, and in the memoirs and reminiscences of those who knew him, some hints can be found. But without surveying the larger context of Cantor's family history, including details of his personal and medical history, it is impossible to appreciate the reasons why set theory, and in particular transfinite set theory, persisted in the face of continuing opposition to thrive in the hands of its creator, Georg Cantor.

*Earlier and somewhat abridged versions of this paper were read to the Section of Mathematics of the New York Academy of Sciences on April 1, 1976, and to the PAREX meeting devoted to "The Social Context of Mathematics: 19th Century Schools of Thought," held at the University of Regensburg, Federal Republic of Germany, July 1–3, 1976. This research has been supported, in part, by the National Science Foundation and by the Faculty Research Foundation of the City University of New York.

†This paper is substantially Chapter XII of Joseph Dauben's book, *Georg Cantor: His Mathematics and Philosophy of the Infinite,* soon to be published by Harvard University Press.

0077–8923/79/0321–0027 $01.75/1 © 1979, NYAS

Georg Cantor: Biography, Genealogy, and Mathematics

What follows is not an attempt to prepare the way for any sort of detailed Freudian psychoanalysis to which Cantor's mind might be submitted. Still, it may as well be said at the beginning that much of Cantor's personality was shaped under the very strong influence of his father. To understand why Cantor approached certain problems and harbored specific anxieties, as he did, it is necessary to sketch what little is known of the family history, and to add some relevant new facts from sources hitherto unpublished.

Of Cantor's grandparents, the family history is unclear, but there is agreement that his grandfather, Jakob Cantor, lived in Copenhagen.[1] Exactly when Cantor's father, Georg Woldemar Cantor was born, is uncertain, though at the time of his death in Heidelberg the official register noted his birth (in Copenhagen) as falling on March 24, 1814. But this does not accord at all with traditional family history, which tells of the bombardment of Copenhagen by the English in 1807.

As a result, the Cantors lost everything and moved to St. Petersburg where Georg Woldemar's mother apparently had relatives. There is, however, evidence for an earlier date of birth: a Danish passport (examined by an expert of the Danish Genealogical Institute) issued to Cantor's father in August of 1833, which stated that he was twenty-four at the time, placing his birth in the year 1809. This fits more accurately with the timing of the English seige of Copenhagen, but not exactly. Nevertheless, it is known that, following the move to St. Petersburg, the upbringing and education of the child Georg Woldemar was entrusted to the Evangelical Mission.[2]

What became of his parents at this time is unknown, though his mother's family, by the name of Meier, was a respected and successful family in St. Petersburg. A sister was married to a chamber musician, Josef Grimm, of the royal court, who was a Catholic. A nephew later went on to become a professor of law at Kazan—as the family liked to recall, this nephew not only taught Tolstoy at one point, but was also instrumental in promoting the legal apparatus involved in freeing the serfs of Russia. As for Georg Woldemar's parents, nothing more is known except for mention in a report of the Danish Genealogical Institute that Jakob Cantor was still alive in 1841, when he sent congratulations to his son on the occasion of his engagement.

Georg Woldemar Cantor married Maria Anna Böhm in St. Petersburg on April 21, 1842.[3] She had been born in St. Petersburg and baptized a Roman Catholic. The entire family was very gifted musically, and included several classical violinists; the most noted was Cantor's great uncle Joseph Böhm, director of the conservatory in Vienna and founder of the great school of violinists which produced many a virtuoso at the time.

The fact that Georg Woldemar and Maria Anna Böhm were married in the Evangelical Lutheran Church in St. Petersburg again reflects the close links that Cantor's father felt for the Mission where he had been raised. All of the couple's six children, of which Georg was the eldest, were baptized as Lutherans. They were given very strict instruction by their father in matters of religion. The seriousness and pervasive consciousness of religion in Georg Woldemar's daily life is strongly reflected in nearly every letter he addressed to his son—letters the young Cantor saved during his school days in Darmstadt and later at the Polytechnic Institute in Zürich. One letter in particular is well known and deserves special mention because it is a model of its kind. The letter, which Cantor's father wrote at the time of his son's confirmation, was one that Cantor never forgot—he kept it with him always. It deserves, despite its length, to be cited nearly *in toto*, for, in so many ways, it seems nearly a fateful prediction of what eventually came to pass as Cantor's personal life and public career began to unfold:[4]

Dearest Georg,

May this day be of blessed influence upon your entire future life through the goodness of the Almighty, the Creator of the universe and Father of all living creatures. May you constantly and unremittingly keep before your eyes the virtuous resolutions which you have no doubt made today in silence as a solemn vow! . . . . The future of one's *life* and the fate of the individual lie hidden from him in the most profound darkness. And it is good that it is so. No one knows beforehand into what unbelievably difficult conditions and occupational circumstances he will fall by chance, against what unforeseen and *unforeseeable* calamities and difficulties he will have to fight in the various situations of life . . . .

Often the most promising individuals are defeated after a tenuous, weak resistance in their first serious struggle following their entry into practical affairs. Their courage broken, they atrophy completely thereafter, and even in the *best* case they will still be nothing more than a so-called ruined genius! . . . . Indeed, it is *not seldom* that young men come to such an end, even those who are *seemingly* endowed with the most promising features of mind and body, and whose prospects for the future through ability and family connections in their youth were also the rosiest in the world!

But they *lacked* a *steady heart* (feste Kern), upon which *everything* depends! Now, my dear son! *Believe me,* your *sincerest, truest,* and *most experienced* friend—this sure heart, *which must live in us,* is: a *true* religious spirit (Gemüt)! This reveals *itself to us* through a *sincere, humble feeling of the most grateful reverence for God,* from which feeling grows as well the *victorious, unshakeable, enduring faith in God,* and which keeps and maintains us throughout our entire life in that silent, undoubting exchange with our heavenly Father! . . .

But in order to prevent as well *all those other* hardships and difficulties which inevitably rise against us (entgegentürmen) through jealousy and slander of open or secret enemies in our eager aspiration for success in the activity of our *own specialty or business;* in order to combat these *with success* one needs above all to acquire and to appropriate the greatest amount possible of the most basic, diverse technical knowledge and skills. Nowadays these are an *absolute necessity* if the industrious and ambitious man does not want to see himself pushed aside by his enemies and forced to stand in the second or third rank.

To the procurement of diverse, thorough, scientific and practical knowledge; to the *perfect* acquisition of foreign languages and literatures; to the many-sided development of the mind in many humanistic disciplines—in order by means of all this first to *equip yourself with dignity* for *those struggles* yet to come—and of this you must always be thoroughly conscious!—*to all this,* the *second period* of your life, your *youth,* now just beginning, is *destined.* Whatever one neglects in *this period* or through *premature* extravagance of his best strength, health, and time, he *squanders* (verludert), so to speak, that is irretrievably and *irreplaceably* lost *for ever;* just as innocence, *once lost,* is for ever and eternally, irretrievably lost . . . .

I close with these words: Your Father, or rather your parents and all other members of the family both in Germany and in Russia and in Denmark have their eyes on *you* as the eldest, and expect you to be nothing *less* than a Theodor Schaeffer and, God willing, later perhaps a *shining star* on the horizon of science.

May God give you *strength,* persistence, health, sound character, and His best blessings! And therefore *you* should follow only on His course. Amen!

Your Father

All of the major features of Cantor's later personality seem etched beneath the words of his father's pen: the religious faith meant to help one through the darkest hours and most troubled times, the certainty that those who failed in life lacked a strong spirit, that such spirit had to reside in the individual as a truly religious soul. To face the slanders of open or of secret enemies and to meet them successfully, one had to be fully prepared, grounded in the expertise of one's profession. The call to devoted and thorough study was driven home with a reminder that the entire family, in Germany, Russia, and Denmark, expected Cantor to be as great as his teacher, Theodor Schaeffer, and perhaps even a renowned man of science.

In large measure Georg Woldemar Cantor's advice to his fifteen year old son reflected his own experience. Raised and educated in an evangelical mission, the father had obviously been deeply impressed by the religious values he had been taught, and was determined to pass them on to his children. As a young man, Georg Woldemar had gone into business in St. Petersburg, and established a wholesaling firm in 1834 under the title of "Cantor and Company." The firm's dealings were international, with interests in Hamburg, Copenhagen, and London; transactions stretched as far as New York, Rio, and Bahia. But for unknown reasons, Cantor and Company folded in December of 1838, and by the end of 1839 liquidation was complete.[5] Nevertheless, Georg Woldemar Cantor still had the "inner resources and determination" to make a success of his life in business, and turned his wholesaling talents to good use as a broker on the stock exchange in St. Petersburg.[6]

His success in business was clearly reflected in his estate. When Georg Woldemar finally died of tuberculosis in June of 1863 he left his family nearly a half million marks. But more than money, he left his eldest son with a compulsive desire to succeed professionally, with a belief in his own inner resources with which he could always be sure of overcoming even the most unpleasant and distressing episodes of life. Cantor was to follow his father's advice with great seriousness. This, as much as any other aspect of his personal life, helps to explain why he was always determined to prevail and why he fought with such tenacity and confidence in support of his theory of sets and transfinite numbers.

Cantor's father took a special interest in his son's education, and was careful to direct both his personal and his intellectual development thoughtfully and with much encouragement. The closeness of their relationship was particularly noticeable in the letters Georg Woldemar sent to his son during the years he was away at school.

After leaving Russia, the family lived shortly in Wiesbaden before settling permanently in Frankfurt. Young Georg, however, boarded at the Realschule in Darmstadt, where he graduated with the report that: "his diligence and fervor are exemplary; his knowledge in basic mathematics including trigonometry is very good; performance praiseworthy."[7] A year later he was at the Höheren Gewerbeschule (trade school) and, again, upon leaving in 1862, was given the highest praise: "He showed himself to be very gifted and a highly industrious pupil."[8] Much to his father's satisfaction, Cantor was working with great industry and concentration, and seemed to have won special recognition for his mathematical abilities. The careful, emphatic admonitions to work diligently and thereby ensure success had not been lost on young Georg.

The fact that Cantor spent two years in the Darmstädter Gewerbeschule suggests that his father's earliest hopes for his son's future did not include the devotion to pure science that his son was finding increasingly hard to resist. But by the spring of 1862 the young Cantor had come to a decision: to devote his life to mathematics. Despite his initial uncertainty as to whether it was the right choice to make, he was wholly supported by his father in the decision. Cantor's joy at his father's approval was echoed in every word of the following letter.[9]

> My Dear Papa,
> You can imagine how very happy your letter made me; it determines my future. The last few days have left me in doubt and uncertainty. I could reach no decision. My sense of duty and my own wishes fought continuously one against the other. Now I am happy when I see that it will no longer distress you if I follow my own feelings in this decision. I hope that you will still be proud of me one day, dear Father, for my soul, my entire being lives in my calling; whatever one wants and is able to do, whatever it is towards which an unknown, secret voice calls him, *that* he will carry through to success!

From the very beginning, Cantor had felt some inner compulsion to study mathematics. Georg Woldemar's original doubts and disappointment over his son's wish to study pure science in university were set aside, but this also left Cantor with a sense of having to prove that his father would not be disappointed. In a move that was later to become characteristic, Cantor reinforced the validity of his decision by saying that it was not his alone. An unknown, secret voice was making its impression even then.

Cantor's father followed his son's career in university with enthusiasm. In August, Cantor took the Reifeprüfung which would qualify him for advanced study in the sciences. Georg Woldemar sent his best wishes, and again stressed a philosophy to which his son would later turn with compelling frequency in periods of doubt and depression. His father's words were a strident call to success, but, as Cantor sat down to write his examination on August 18, 1862, he had more than his father's prayer that God might grant him blessings and luck. He also had his father's reminder (which he would carry through life):

> May God give you His blessing and good fortune for the examinations beginning today. . . . In critical moments of one's life an unwavering and joyous trust in God and a profound, moving prayer to the Almighty Giver of all goodness before the start of the day gives stability, courage, and self-confidence. And thus commit yourself to God! Be fresh, joyous and cheerful in work![10]

By the end of August the results of the examination had been tabulated, with Cantor's performance meriting the comment: "Recognized as proof of first-level preparation (Maturitäts Zeugnis Nr. 1) for the study of the natural sciences."[11] Apparently Cantor had done less well in subjects like geography and history, which prompted his father to stress the importance of a diverse and thorough study of all branches of knowledge.[12]

Before Cantor left Darmstadt, his father had counseled that one should not merely concentrate on a single subject, like botany, but that young Georg should study everything from languages to music as thoroughly as practicable. Thus he was delighted to hear of his son's interest in hearing a series of lectures by the renowned Vischers on Shakespeare, of which his father wrote enviously that he had never heard anything but the highest praise for the man.

When Cantor formed his own string quartet, his family was delighted, and invited the entire group to Frankfurt for the Christmas holidays. But nothing indicates so well the amount of paternal pride invested in his son than the solemnity and seriousness with which Georg Woldemar wrote to his son about a drawing Cantor had sent home as evidence of the progress he was making at the Polytechnicum in Zürich.

> Whereas Georg Ferd. Louis Phil. Cantor has not spent years in the study of drawing according to the ancient models; and whereas this is his first work and in this difficult art a perfected technique is only achieved after great diligence; and whereas, futhermore, until now he has greatly neglected this beautiful art; the thanks of the nation—I mean of the family—is unanimously voted to him for this first effort, which already shows great promise.[13]

Contrary to the outrageous claims made by E.T. Bell[14] that Georg Woldemar had a thoroughly deleterious and ruinous effect upon his son's psychological health, the surviving evidence of their relationship indicates the exact opposite. Cantor's father was clearly a sensitive and gifted man, who loved his children deeply and wanted them to live happy, successful, and rewarding lives. Being thoroughly religious, he passed similar and deeply felt convictions on to his first son, Georg, in full measure.

As a child, young Georg was expected to be a credit to his family, and, as he approached a career in mathematics, he wanted nothing more than to fulfill the best hopes of his parents. Rather than disappoint them, he worked diligently and proved to be an energetic student. But underlying even his earliest decisions was a sense of destiny, a feeling that unknown forces were at work which pulled him towards a career he felt he could not deny.

Personality, Psychology, and Depression

Cantor's first serious mental breakdown, occurring shortly after his thirty-ninth birthday in the spring of 1884, has always figured prominently in the accounts of his life given by more sensational writers of the history of mathematics. Of these, none has treated his case with less reliability than E.T. Bell.[15] Not content with the view advanced by Schoenflies[16] in 1927 that Cantor's breakdown was the result of conflict with Kronecker and his inability to solve the continuum hypothesis, Bell turned to the popular psychology of his day and produced a predictably Freudian analysis. He traced the roots of Cantor's mental instability back to childhood and blamed Cantor's father for having left his son with deeply planted anxieties that eventually produced psychological instability requiring hospitalization. But unless Bell had access to material (never cited) unknown to historians before or since, there is no basis at all for the distorted scenario he paints with such gusto. On the contrary, the actual details of Cantor's illness, as it is currently understood, are considerably more interesting than the maudlin Freudian analysis Bell conjures, and, in fact, make a great deal of sense when keyed to other features of Cantor's personality.

When he suddenly suffered his first breakdown in May of 1884, Cantor had just returned from an apparently successful, quite enjoyable trip to Paris.[17] He had met a number of the French mathematicians, including Hermite, Picard, and Appell, and was delighted to report to Mittag-Leffler that he had liked Poincaré very much and was happy to see that the Frenchman understood transfinite set theory and its applications in functional analysis.

Since most French mathematicians were away from Paris on vacation, Cantor had a good deal of leisure to pursue his own work, and spent the remainder of his time visiting the galleries and museums, indulging his love of music with two visits to the opera and including an evening of theater at the Comédie Française.

But unexpectedly, after eight days, he was called back to Germany to deal with family matters of some sort in Frankfurt, and shortly thereafter, his breakdown occurred. It is impossible to conjecture the immediate cause, though one theory, made popular by Schoenflies,[18] blames the collapse on Kronecker and the intractability of the continuum hypothesis.

Cantor's bitterness, the product of tremendous opposition to his work, coupled with his inability to produce the most important result of all, drove him to the heights of doubt and distress, eventually causing him to despair of ever making a success of his mathematics.

His efforts to solve the continuum conjecture reached a fervent pitch. Determination drove him on, and as his physical stamina was strained, the breakdown inevitably followed. He lapsed into the abyss of depression which left him weak and unsure of himself and of his transfinite set theory.

Schoenflies took Cantor's attempt to reconcile matters with Kronecker later that summer as proof that this reconstruction of the causes and events leading up to the crucial moment of collapse was essentially correct.[19] Cantor himself recalled the entire episode in very similar terms when he wrote to his friend, Mittag-Leffler, a few months later on August 18, 1884.[20] Cantor was in Friedrichroda at the time, where he was trying to relax and to recuperate from the entire unhappy experience. He felt that it had all been his own fault, brought on by problems with his research.

The breakdown had not been caused by overwork, he insisted, but by irritations and frictions with Kronecker, for one. He realized he should never have let Kronecker upset him to such a degree. Consequently, the best cure was to face Kronecker squarely and to attempt a reconciliation, which is exactly what Cantor did. On the same day he wrote to Mittag-Leffler, he posted a letter to Kronecker and tried to explain his unhappiness over the state of their rivalry.[21]

Above all, he hoped it was still possible to gain some semblance of equanimity. Whatever might come of the attempt, Cantor knew that writing to his rival would at least serve to lighten his heart. By the end of August Kronecker had written a polite, even friendly response, reminding Cantor of how much closer they had been when he was a student in Berlin, and registering some surprise at Cantor's sudden claims to any hostility or animosity between them.[22] Cantor was greatly reassured, and wrote off immediately to Mittag-Leffler to say how happy he was, that at least he could work again without the dread of personal animosity.[23]

In retrospect, the first major breakdown was relatively short, and could not have lasted more than a month. A letter to Felix Klein of May 10, 1884 was apparently the last before onset of the illness. On June 21 he wrote again to Mittag-Leffler saying he had not felt so "fresh" lately and could not be sure when he would get back to his research.[24]

Despite the speed with which the depression seemed to pass, it left the entire family shaken. His eldest daughter Else, only nine at the time, was bewildered by the incomprehensible change in her father, the swiftness with which his entire manner had been transformed.[25] Cantor himself emerged from his attack emotionally enervated, and the following months were spent in trying to rebuild his strength and to regain his intellectual spontaneity.

Though he returned to mathematics that fall and again attempted to work on the continuum hypothesis, his attitudes generally had undergone substantial alteration. He began to emphasize other interests. The amount of time he devoted to various literary-historical problems steadily increased, and he read the history and documents of the Elizabethans with great attentiveness, hoping to prove that Francis Bacon was the true author of Shakespeare's plays.

As time progressed, he also began to deepen his study of scripture and the church fathers, and developed interests in matters of Freemasonry, Rosicrucianism, and Theosophy.[26] Perhaps he felt that his earlier exclusive devotion to mathematics had been too concentrated, too painful, without the rewards he had expected to compensate and make it all worthwhile. Perhaps he hoped to dilute the intensity of his earlier devotion to mathematics and to take life more easily, thereby avoiding the chances of a relapse of his illness.

Nothing reflects more directly Cantor's shift in attitude following his first breakdown than does his sudden desire to teach philosophy in Halle instead of mathematics.[27] Apparently he was striving for a certain balance as he began to take his extramathematical interests more and more seriously.

The strategy was successful for about fifteen years. But then the breakdowns grew more intense, lasted longer, and occurred with greater frequency. The year 1899 brought with it a startling series of events. It was the first year since 1884 in which any record survives of his hospitalization for reasons of mental instability.[28]

During the summer the antinomies of set theory were on his mind, as we know from the letters he sent to his friend Dedekind.[29] But Cantor was certainly unhappy at the unsatisfactory impasse he had reached in his old problems of the continuum hypothesis, well-ordered sets, and the comparability of the transfinite numbers.

He applied for a leave of absence from university for the fall term, and his request was granted. But in November he sent an extraordinary letter to the Ministry of Culture, saying that he was anxious to abandon completely his position at the university.[30] So long as his salary was not diminished (he asked for no increase from his recompense at university of 6000 marks annually), he would be content with a modest position in some library. Nor was a prominent title necessary; he was ready to accept virtually any alternative that might release him from the confines of the German university.

Whether or not the Ministry could cooperate, he was determined to break away "by all means." He emphasized his qualifications, his knowledge of history and literature, his publications on the Bacon-Shakespeare question, and even added the provocative news that he had come upon certain information in the course of his research concerning the first king of England, "which will not fail to terrify the English government as soon as the matter is published."[31] Cantor tried to inject a note of urgency into the matter by asking the ministry to send him their reply within the next two days, for, if they could offer him no alternative to teaching, then as a born Russian he would apply to the Russian diplomatic corps in hopes that he might be of service to Czar Nicholas II.[32]

Nothing seems to have come of Cantor's request, nor did he enter the ranks of the Russian diplomatic service. The entire episode fit the pattern of his continuing fears of persecution and opposition, which previously, in 1884, had prompted him to think seriously about giving up mathematics in order to teach philosophy. But he was no more successful in 1899 than he had been in 1884. Finding neither a post as librarian nor a position in diplomatic service in Berlin, Cantor remained at the university in Halle. For part of 1899 he was again in hospital, but before the year was out, a final tragic event occurred which produced a surpassing disillusionment and much pain for the entire family.

Cantor's mother was born in St. Petersburg in 1819 and died in 1896 in Berlin, more than thirty years after the death of her husband. Three years later her son, Constantin, Cantor's younger brother, died in Capri, where he had retired from military life as an officer in a Hessian Dragoon Regiment to marry an Italian Baroness.[33] Then, on December 16, 1899, while Cantor was delivering a special lecture on the Bacon-Shakespeare question in Leipzig, his youngest son died suddenly. Cantor did not return home until later that evening to discover the terrible news: Rudolf had died towards midafternoon. In just four days the child would have been thirteen.

Cantor described little Rudolf in a letter to Felix Klein a week later and the tragedy was borne in every word he wrote.[34] The boy had been frail in childhood, but then began to grow stronger. His spirit correspondingly strengthened, he was so loving, so amiable, that he was everyone's favorite. Rudolf had been gifted musically, and his father hoped that he might follow in the family tradition and become a famous violinist. In writing to Klein, Cantor was prompted to recall his own early study of the violin and to rekindle his doubts about the wisdom of having given up his inherited talents and love of music to become a mathematician.

His early enthusiasm for science had faded from memory; he could not even remember why he had ever given up music for mathematics.[35] That early voice calling him to a new and challenging life in 1863 seemed entirely forgotten and buried under the weight of events since. In the disappointment and unhappiness which his own choice of profession seemed to have produced, Cantor wanted something better for his son. Rudolf represented a means through which he might still live and enjoy a way of life he now regretted having abandoned. But now, even that hope was gone.

Despite his hospitalization and the tragic events of 1899, Cantor managed to maintain his equilibrium for another three years, until he was again hospitalized and relieved of his teaching duties for the winter term 1902–1903.[36] Barely a year later another shock sent him to the hospital again, and thereafter he was frequently in and out of the Halle Nervenklinik.

The dramatic events of Jules König's paper read during the Third International Congress for Mathematicians in Heidelberg greatly upset him, as we know.[37] He was there with his two daughters, Else and Anna-Marie, and was outraged at the humiliation he felt he had been made to suffer. He seemed less upset at the suggestion his theory was somehow wrong, for that was impossible, unthinkable. Instead, he was furious at the public way in which his work had been brought to question.

As the Congress drew to a close, he was still in a state of agitation, and Schoenflies' story[38] of how Cantor boomed into breakfast a week later to describe in excited terms the error in König's proof, shows that his mind was racing to find ways of dispelling the doubts that lingered in the aftermath of the Congress.

It is no surprise that the strain was too much for him; he was soon again hospitalized and given a leave from teaching for the winter term 1904–1905. Thereafter he spent increasingly longer periods at the Nervenklinik in Halle, from October 1907 to June 1908, and from September 1911 to June 1912, when he was then moved to yet another sanatorium.[39] He was admitted to the Halle clinic for the last time on May 11, 1917. By then the end was near. He did not want to go and wrote continuously to his family asking that they come to take him home.[40]

Summer passed to autumn, but Cantor was not allowed to return home. World War I was still raging and food was scarce. The lack of nourishment shows in a surviving photograph—Cantor's face is gaunt and tired, his eyes while still a sparkling blue were no longer piercing, no longer full of questions; instead they were passive and reflective. On January 6, 1918, he died apparently of heart failure. Edmund Landau wrote to Cantor's wife as soon as he had heard the news, saying that Cantor and all that he represented would never die. One had to be thankful, he observed, that humanity had been given a Georg Cantor from whose works later generations would learn: "Never will anyone remain more alive."[41]

The Significance of Cantor's Nervous Breakdowns

With the above details of Cantor's biography and his medical history in mind, it is now appropriate to consider their significance. Above all, there is reason to suspect that much more is involved in appreciating the nature and causes underlying Cantor's periods of euphoria and depression than Schoenflies' conventional interpretation might suggest.[42] After all, the account which Schoenflies published was concerned exclusively with Cantor's first major breakdown of 1884, and it was not difficult for him to draw explicit lines between Cantor's illness and specific anxieties which the climate of Cantor's research had produced.

But what of the later episodes? Though, at points, particularly in 1904, there seem to be links between periods of emotional upset and difficult impasses with his mathematics, it is by no means clear that the troubles with research and professional rivalries were the sole cause of his breakdowns.

The traditional historiography, passed by word of mouth from mathematicians of Cantor's own generation to the next, was eventually codified in Schoenflies' account of Cantor's "critical years." The theory was elaborated, some slicker elements added, packaged anew, and given great currency through the popular writings of E.T. Bell.[43]

A final, even more unfortunate stamp of authenticity, considering the source, came from Bertrand Russell, who shortly before his own death wrote in somewhat piquant terms that "Georg Cantor, the subject of the following letter, was in my opinion, one of the greatest intellects of the nineteenth century. . . . After reading the following letter, no one will be surprised to learn that he spent a large part of his life in a lunatic asylum."[44]

Fortunately, we are now in a somewhat better position to assess the nature and consequence of Cantor's nervous breakdowns. Ivor Grattan-Guinness was the first to consider Cantor's illness from a clinical point of view by reviewing what records survive and reconstructing the outline of a case history: "The attacks all began suddenly, usually in an autumn season, and exhibited phases of excitement and exaltation; they ended suddenly in the following spring or summer, and were sometimes followed by what we now understand to be the depressive phase. In Cantor's day this would have been seen as a cure, and he would be sent home to sit silent and motionless for hours on end." He then concluded, with the advice of a psychologist who examined Cantor's file in the Halle Nervenklinik, that "Cantor's illness was basically *endogenous,* and probably showed some form of manic depression: exogenous factors, such as the difficulties of his researches and the controversies in Halle University, are likely to have played only a small part in the genesis of his attacks, little more than the clap that starts the avalanche. Thus he would have suffered his attacks if he had pursued only an ordinary mundane career."[45]

Moreover, there is good evidence of a direct nature which tends to confirm that this diagnosis is more realistic and in accord with the known facts of Cantor's history than any previous attempts to explain his mental illness. For example, well before his first nervous breakdown in 1884, Cantor had been prone to periods of depression, and such episodes help to support the view that the origins of his mental illness were not determined by his anxieties over mathematical doubts or rivalries.

In 1863 when Cantor was beginning his studies at the Polytechnicum in Zürich, something happened to promote serious misgivings about his future, and to a drastic degree. His father sent off a worried letter to his son on January 21 in an attempt to help him out of the doldrums. The letter began with a little quatrain, somewhat frivolous, but designed to coax Cantor from his moodiness.[46] Then followed a number of exhortations that might just as easily have applied to Cantor's later bouts of depression. Above all, Cantor's father wanted to dispel young Georg's melancholy. If only he were home, his father was sure he could break his son's depression, cheer his state of mind to such a degree that he would be immune to any relapse for three years, at least!

Was this an implication that such moods had happened before? Or was his father only saying that he could liven Cantor's spirits for some time were he not so far away? Whichever was true, Cantor's father realized that his son's moodiness was the result of some unfounded anxiety. It caused the young student to work long into the night, endangering his health; but even more, it prefigured the patterns associated with his later periods of fatigue and emotional disturbance.

Apparently Cantor was anxious about the costs of his education, and he doubted his ability to succeed were he forced to hurry his studies and face examinations before he was fully prepared.[47] Such anxieties compelled him to study without interruption. He even went without sleep if necessary. This left him tired and eventually so despondent that his father was worried about his condition.

Georg Woldemar wrote reassuringly, and made it clear that means were not lacking to allow his son to take as much time as might be necessary to complete his studies. But within five months his father would be dead of tuberculosis, and Cantor would have to leave Zürich. Moreover, he would then be without the positive and understanding support which his father had always given. Cantor had promised to make a success of his career in mathematics, but after June, 1863, with the death of his father, he was left to do so alone.

Cantor's Muse: The Inner Voice and Divine Inspiration

Too little information has been preserved to allow any detailed assessment of Cantor's personality, which leaves the historian to say either nothing on the subject or to conjecture as best he can. With this caveat in mind, the analysis which follows can only plead reasonableness by virtue of its consonance with the known facts of Cantor's history; it makes no claims to being the only reading possible.

On September 17, 1904, Cantor was admitted to the Halle Nervenklinik, where he remained until March 1, 1905. When he was released, he did something quite remarkable, something which, I believe, throws an unsuspected light upon his periods of mental breakdown. At least in Cantor's own mind his periods of depression and seclusion served a unique and generative purpose, providing rest and quiet during which great progress might be made in many facets of study by mere reflection.

The most surprising declaration Cantor made upon his release from hospital in 1905 was that he had had an "inspiration from above, which suggested to me a renewed study of our *Bible* with opened eyes and with banishment of all previous preconceptions."[48] He was writing to the English mathematician, P.E.B. Jourdain, and explained how the "captivity and solitude" of his confinement had produced unexpected insights from which Cantor produced his pamphlet *Ex Oriente Lux*.[49] It was written as a series of dialogues between a master and his pupil, and the major point of the piece was to argue that Christ was the natural son of Joseph of Arimathea!

The point of all this is not to show that Cantor was interested in somewhat peculiar aspects of "documentary Christianity,"[50] as he referred to the pamphlet *Ex Oriente Lux,* but rather to show how he regarded his periods of hospitalization. During the long months of seclusion his mind was left free to ponder many things, and in the silence he could perceive the workings of a divine muse—he could hear a secret voice from above which brought him both inspiration and enlightenment. Moreover, these convictions also provided a similar link between Cantor's periods of introverted contemplation, the long silences, and his mathematics.

Following another long period of hospitalization at the Halle Nervenklinik in 1908, Cantor wrote to his friend in Göttingen, the English mathematician Grace Chisholm Young. Apparently, Cantor's inner voice knew more than the details of Christian history—his muse was also a mathematician! Here it is best to let Cantor speak for himself:

> A peculiar fate, which thank goodness has in no way broken me, but in fact has made me stronger inwardly, happier and more expectantly joyful than I have been in a couple of years, has kept me far from home—I can also say far from the world—since October 22, 1907, until June 15, 1908, thus until last Monday. . . . In my lengthy isolation neither mathematics nor in particular the theory of transfinite numbers has slept or lain fallow in me; the first publication in years which I shall have to make in this area is designated for the "Proceedings of the London Mathematical Society," to which, for the great honor of having named me to its membership, I shall always remain grateful as well as to the Royal Society for conferring upon me the Sylvester Medal three and one-half years ago.[51]

At the close of this same letter he lashed out against Poincaré's remarks condemning Cantorism at the International Congress in Rome, 1908. Poincaré believed that mathematics could be saved from the paradoxes of set theory by allowing only those concepts which could be completely defined in a *finite* number of words. Instead of generalizing, building more complicated theorems from simpler, established results, Poincaré criticized Cantor for starting from the *Genus supremum,* or the assumed existence of the actual infinite.

In his letter to Mrs. Young, Cantor made reference to the impossibility of any sort of *Genus supremum* embracing *all* transfinite numbers. In fact, he was adamant in underscoring that:

> I have never proceeded from any "Genus supremum" of the actual infinite. Quite the contrary, I have rigorously proven that there is absolutely no "Genus supremum" of the actual infinite. What surpasses all that is finite and transfinite is no "Genus"; it is the single, completely individual unity in which everything is included, which includes the "Absolute," incomprehensible to the human understanding. This is the "Actus Purissimus" which by many is called "God."[52]

There can be no mistake about Cantor's identification of his mathematics with some greater absolute unity in God. This also paralleled his identification of transfinite set theory with divine inspiration. Even *before* his first nervous breakdown of 1884, Cantor had told Mittag-Leffler that his transfinite numbers had been communicated to him from a "more powerful energy"; that he was only the means by which set theory might be made known. Just as he had been inspired to write *Ex Oriente Lux,* a "Dialogue of a master with his student . . . reported by the student himself," so too his mathematics could be inspired through a muse from above.

Thus the periods of isolation in hospital could be regarded as periods during which, as he told Mrs. Young, the transfinite numbers lay neither fallow nor forgotten, but might be further elucidated by the grace of God sent to inspire new lines of research. All this was very much in keeping with the principles Cantor had inherited from his father and from his religious upbringing.

Georg Woldemar had been careful to instill in his son a reverence and faith in one's ability to succeed by hard work and faith in God. Without such confidence in his own abilities, Cantor might never have had the courage to face the relentless opposition he encountered to his work almost from the start. Had he not been able to cast himself in the role of God's messenger, inspired from some higher source of inspiration, he might never have asserted the unquestionable, indubitable correctness of his research. The religious dimension which Cantor attributed to the *Transfinitum* should not be discounted as merely an aberration. Nor should it be forgotten or separated from his existence as a mathematician.

The theological side of Cantor's set theory, though perhaps irrelevant for understanding its mathematical content, it nevertheless essential for the full understanding of his theory and why it developed in its early stages as it did. Cantor believed that God endowed the transfinite numbers with a reality making them very special. Despite all the opposition and misgivings of mathematicians, in Germany and elsewhere, he would never be persuaded that his results could ever be imperfect. This belief in the absolute and necessary truth of his theory was doubtless an asset, but it also constituted for Cantor an imperative of sorts. He could not allow the likes of Kronecker to beat him down, to quiet him forever. He felt a duty to keep on, in the face of all adversity, to bring the insights he had been given as God's messenger to mathematicians everywhere.

THE EARLY DEVELOPMENT OF TRANSFINITE SET THEORY: THE EFFECT AND INFLUENCE OF CANTOR'S CONTRIBUTION

One has only to read the opening sections of Cantor's *Grundlagen* to see how his interest in philosophy affected his presentation of set theory. We know that he went into great detail concerning the philosophy of the infinite because he wanted to confront and dispel the long-standing objections traditionally levied against the idea of actual infinities. Cantor knew his work was revolutionary; he too had found it difficult to overcome long-standing prejudice against completed infinities.[53] But gradually he came to see the power of his new ideas and he found that it was possible to dispel even the objections of philosophers and theologians. Nevertheless, his heavy philosophical encasement of set theory in the *Grundlagen* was a disadvantage to its widespread acceptance. This was suggested by Felix Klein's appeal and mentioned specifically in Mittag-Leffler's plea that Cantor remove all such nonmathematical material from his publications on the subject.[54]

Cantor, however, continued to believe that his mathematics and his metaphysics went hand in hand. As a result, many thought of set theory as belonging more to philosophy than to serious mathematics, and for many years journals and mathematical indices alike continued to list works in set theory under the rubric "Philosophy."[55] Only after publication of the *Beiträge,* completely mathematical and devoid of any overtly philosophical content, did his theory of transfinite numbers begin to receive its full share of study and recognition.

If philosophy itself affected the form and early acceptability of Cantor's set theory, his deeply religious convictions played a rather different role. His personality, as we have considered it in some detail, was such that success was very important to him, and came almost as a promise to prove to his parents that his decision to study mathematics had been the proper one.

But, as the years passed, theoretical difficulties with the continuum hypothesis and mounting external criticism from opponents like Kronecker conspired to raise doubts and to feed Cantor's anxieties about the acceptability and success of his work.

As a result of his breakdown in 1884, he was forced to deal with the problems that had provoked his illness. He attempted a reconciliation with Kronecker, but that was short-lived. He worked less fervidly at his mathematics and, for a time, seemed rather detached from any interest in the success of promotion of set theory. What seemed to keep his spirit alive, however, was a growing faith in the absolute certainty of his theory of transfinite sets.

Perhaps a weaker man would never have recovered from the experiences of 1884–1885. It is conceivable that others might have given up so seemingly hopeless a struggle against criticism of such power and influence as Kronecker's, but Cantor refused to be intimidated. He could count a special resource of his own that was immensely significant to the fortunes of his mathematics.

Because he believed that set theory, having been divinely inspired from God, was therefore absolutely and necessarily true, nothing could succeed in shaking his confidence that it could never be flawed. It was exactly this point that he once made in a letter of 1888 to Heman: he knew his theory was correct because he had taken care to investigate over the years all objections that could conceivably be raised against it. In every case Cantor was satisfied that he could defeat any opposition, for the simple reason that:

> My theory stands as firm as a rock; every arrow directed against it will return quickly to its archer. How do I know this? Because I have studied it from all sides for many years; because I have examined all objections which have ever been made against the infinite numbers; and above all, because I have followed its roots, so to speak, to the first infallible cause of all created things.[56]

Such unwavering, infallible convictions were Cantor's greatest allies in periods of stress and difficulty. As his father had predicted, faith in God and a certainty in the capacity of one's own abilities to meet any problem, however great, would always ensure that one could weather any storm.

Had it not been for the force of his own faith, supported by convictions entirely nonmathematical in nature, his theory of transfinite numbers might never have survived the criticism of finitists and intuitionists alike. Cantor did much more for set theory than merely discover its basic principles. He not only shaped its early character and formulated all of its most essential elements virtually single-handedly, but he also ensured that once mathematicians were ready to consider the significance of the transfinite numbers, the entire theory would be ready to stand on the foundations which he had given it.

Though distasteful and ultimately injurious to Cantor's health and well-being, the continuing conflict with Kronecker had its positive results. It forced Cantor to look more carefully at the foundational aspects of the new theory he was building. It caused him to search for more direct, mathematically acceptable, rigorous means by which to introduce, define, and explain transfinite numbers. It might well have prompted him to take all philosophical objections more seriously. And, in the end, it stimulated his concern for the *Deutsche Mathematiker-Vereinigung,* and his earnest desire to promote not only a German forum for mathematicians, but to further, as well, the organization of mathematical congresses on an international scale. Finally, it may be that without the keen and relentless opposition from Kronecker, Cantor might never have been driven to produce such a detailed, carefully argued exposition of his theory as the *Beiträge* of 1895 and 1897.

Later generations might forget the philosophy, smile at the abundant references to St. Thomas and the church fathers, overlook his metaphysical pronouncements and miss entirely the deeply religious roots of Cantor's later faith in the veracity of his work. But these all contributed to Cantor's resolve not to abandon his transfinite numbers for less controversial and more acceptable interests. Instead, his determination seems actually to have been strengthened in the face of opposition. His forbearance, as much as anything else he might have contributed, ensured that set theory would survive the early years of doubt and denunciation to flourish eventually as a vigorous, revolutionary force in scientific thought of the twentieth century.

Notes and References

All citations are keyed by author and date of publication. Unless otherwise noted, all references to the works of Georg Cantor are taken from the edition by E. Zermelo, G. Cantor (1932): *Gesammelte Abhandlungen mathematischen und philosophischen Inhalts* (Berlin: J. Springer, 1932; reprinted Hildeshiem: Olms, 1966). Nevertheless, Cantor's papers are designated by their original years of publication. Thus "Cantor (1872), 93" refers to Cantor's article of 1872 *as given in Zermelo's edition* of the *Gesammelte Abhandlungen* on page 93. For the sake of economy, the following abbreviations are used in references to the three surviving *Briefbüchern* in which Cantor drafted his letters before writing out final copies:

Cantor (I): Cantor's letter-book for 1884 through 1888.

Cantor (II): Cantor's letter-book for 1890 through 1895.

Cantor (III): Cantor's letter-book for 1895 and 1896.

These letter-books are part of Cantor's surviving *Nachlass* which is now kept in the Archives of the Akademie der Wissenschaften, Göttingen. I am grateful to Oberstudienrat Wilhelm Stahl for allowing me access to these materials before they were given to the Akademie.

1. For biographical details of Cantor's family history, refer to A. FRAENKEL (1930): "Georg Cantor," Jahresber. Dtsch. Math. Ver. **39:** 189–266. This biography also appeared separately as *Georg Cantor* (Leipzig: B. G. Teubner, 1930), and is reprinted in an abridged version in CANTOR (1932), 452–483; M. PETERS: *Lied eines Lebens* (a biography of Else Cantor, privately printed: Halle, 1961); H. MESCHKOWSKI (1967): *Probleme des Unendlichen. Werk und Leben Georg Cantors* (Braunschweig: Vieweg); and I. GRATTAN-GUINNESS (1971a): "Towards a biography of Georg Cantor." Ann. Sci. 1971. **27:** 351–352.
2. From the report of the Danish Genealogical Institute, prepared in 1937 by T. Hauch-Tausböll. In *Nachlass Cantor I,* as listed in GRATTAN-GUINNESS (1971a), 348.
3. Cantor related much of his family history in the course of his correspondence. The information recounted below is gathered from numerous letters to be found throughout CANTOR (I), (II), and (III). See in particular the biographical details Cantor wrote out on visiting cards in 1899, transcribed in GRATTAN-GUINNESS (1971a), 379–380.
4. Reproduced in FRAENKEL (1930), 191–192. In the next to last paragraph, however, Fraenkel transcribes "ein leuchtendes *Gestirn* am Horizonte der *Ingenieure,*" while the typescript version in *Nachlass Cantor I* reads "Wissenschaft" instead of "Ingenieure." Part of the confirmation letter is also reproduced in MESCHKOWSKI (1967), 3.
5. From a German translation of the Cantor's family history, prepared by T. Hauch-Tausböll of the Danish Genealogical Institute, Copenhagen, and preserved in *Nachlass Cantor I.* See GRATTAN-GUINNESS (1971a), 351.
6. GRATTAN-GUINNESS (1971a), 352.
7. FRAENKEL (1930), 192.
8. FRAENKEL (1930), 192.
9. FRAENKEL (1930), 193; quoted in MESCHKOWSKI (1967), 5.
10. Georg Woldemar Cantor to Georg Cantor, August 18, 1862, in *Nachlass Cantor I.*
11. FRAENKEL (1930), 193.
12. See Georg Woldemar Cantor to Georg Cantor, August 26, 1862, and May 9, 1861, in *Nachlass Cantor I.*
13. Georg Woldemar Cantor to Georg Cantor, November 3, 1862, in *Nachlass Cantor I.*
14. BELL, E. T. (1937): *Men of Mathematics* (New York: Simon and Schuster), Chapter 29.
15. BELL (1937), Chapter 29.
16. SCHOENFLIES, A. 1927. "Die Krisis in Cantor's mathematischen Schaffen," Acta Mathematica. **50:** 1–23.
17. For literature dealing with Cantor's nervous breakdowns, see SCHOENFLIES (1927); FRAENKEL (1930), 198, 207–210; PETERS 1961; GRATTAN-GUINNESS (1971a), 355–358, 368–369.

    Cantor mentioned his trip to Paris, shortly before his breakdown in 1884, in two letters: Cantor to Mittag-Leffler, May 4, 1884, Letter #18, Archives of the Institut Mittag-Leffler, Djursholm, Sweden; Cantor to Klein, May 10, 1884, Letter #437, Archives of the Universitätsbibliothek, Göttingen.
18. SCHOENFLIES (1927).
19. SCHOENFLIES (1927).
20. Cantor to Mittag-Leffler, August 18, 1884, Letter #22, Archives of the Institut Mittag-Leffler, Djursholm, Sweden.
21. See Cantor's letter to Mittag-Leffler of August 18, 1884.
22. Kronecker to Cantor, August 21, 1884, transcribed in MESCHKOWSKI (1967), 237–239.
23. Cantor to Mittag-Leffler, August 26, 1884, printed in MESCHKOWSKI (1967), 242–243.
24. Cantor to Klein, May 10, 1884, cited in note 17 above; Cantor to Mittag-Leffler, June 21, 1884, Letter #19, Archives of the Institut Mittag-Leffler. See SCHOENFLIES (1927), 9; GRATTAN-GUINNESS (1971a), 376–377.
25. PETERS (1961), 15, 27.
26. For details concerning the amount of correspondence Cantor devoted to such matters, consult for example Cantor's letters to Gutberlet, January 24, 1886, and February 6,

1887, both in CANTOR (I), 36–38, 94-95, resp.; Cantor to Schmidt, March 26, 1887, in CANTOR (I), 98–102; Cantor to Grafen Vitzthum von Eckstadt, August 29, 1887, in CANTOR (I), 104; Cantor to Keipewetter, September 9, 1891, in CANTOR (II), 79–80; and Cantor to Hermite, November 30, 1895, and February 11, 1896, both in CANTOR (III), 45–50, 144–145. Note also Cantor's work on the Goldbach Theorem, discussed in MESCHKOWSKI (1967), 168–172; GRATTAN-GUINNESS (1971a), 360–361; Cantor's letter to Hermite, November 30, 1895, transcribed in part in MESCHKOWSKI (1967), 262–263.

27. Cantor to Mittag-Leffler, October 20, 1884. In this letter Cantor first indicated his desire to abandon mathematics and teach philosophy. Mittag-Leffler wrote back on November 2, 1884, endorsing the idea but adding his hope that Cantor would continue to be the great mathematical *author*. That Cantor actually began to teach philosophy at Halle is clear from Sophie Kowalewski's letter to Mittag-Leffler of May 21, 1885, Letter #35, Archives of the Institut Mittag-Leffler. Cantor wrote to Hermite on January 22, 1894, explaining that for nearly twenty years mathematics had not been his sole interest, but that: "Metaphysik and Theologie haben, ich will es bekennen, meine Seele in solchem Grade ergriffen, dass ich verhältnismässig wenig Zeit für meine *erste Flamme* übrig habe," in CANTOR (II), 126; transcribed in H. MESCHKOWSKI (1965): "Aus den Briefbüchern Georg Cantors," Archive for History of Exact Sciences **2:** 514.
28. GRATTAN-GUINNESS (1971a), 368–369.
29. Zermelo included these letters from Cantor to Dedekind in his edition of the *Gesammelte Abhandlungen:* CANTOR (1932), 443–450. Comparison with the originals shows that Zermelo did a very poor, sometimes surprisingly inaccurate, job of transcribing them. For detailed analysis of corrections that ought to be made in Zermelo's edition, consult I. GRATTAN-GUINNESS (1974): "The Rediscovery of the Cantor-Dedekind Correspondence," Jahresber. Dtsch. Math.-Ver. **76:** 134–136.
30. Cantor to Graf, November 10, 1899: transcribed in GRATTAN-GUINNESS (1971a), 378–379.
31. Cantor to Graf, November 10, 1899.
32. See Document VI of GRATTAN-GUINNESS (1971a), 380–381.
33. GRATTAN-GUINNESS (1971a), 380.
34. "Unser jüngster Kind Rudolf ist uns vier Tage vor seinem dreizehnten Geburtstage, auf dem Wege in einen Handarbeitsunterricht, dem er seit einiger Zeit besuchte, am 16*ten* Dec. durch Herzschlag genommen worden," in Cantor to Klein, December 31, 1899; Letter #455, Archives of the Universitätsbibliothek, Göttingen; transcribed in GRATTAN-GUINNESS (1971a), 381.
35. Cantor to Klein, December 31, 1899.
36. GRATTAN-GUINNESS (1971a), 368–370.
37. Turn to the discussion in G. KOWALEWSKI (1950): *Bestand und Wandel* (München, 1950).
38. For details, refer to A. SCHOENFLIES (1928): "Georg Cantor," *Mitteldeutsche Lebensbilder 3* (Magdeburg: Selbstverlag der Historischen Kommission, 1928).
39. GRATTAN-GUINNESS (1971a), 368. From time to time, Cantor apparently visited other sanitaria, though no official records are known by which exact dates of admission or duration of stay may be documented for periods other than those listed here.
40. CANTOR to Vally Guttmann Cantor, letters in *Nachlass Cantor II,* as listed in GRATTAN-GUINNESS (1971a), 348.
41. E. Landau to Vally Guttmann Cantor, January 8, 1818, in *Nachlass Cantor II;* transcribed in MESCHKOWSKI (1967), 270.
42. SCHOENFLIES (1927).
43. BELL (1937), Chapter 29.
44. RUSSELL, B. (1967): *The Autobiography of Bertrand Russell* (New York: Bantam, 1968), 217.
45. GRATTAN-GUINNESS (1971a), 368–369.
46. Georg Woldemar Cantor to Georg Cantor, January 21, 1863, *Nachlass Cantor I;* cited in MESCHKOWSKI (1967), 2: "Grillen sind mir böse Gäste/immer mit leichtem Sinn/tanzen durch' Leben hin, das nur ist Hochgewinn!" Note that "Grillen" may refer both to crickets and to a melancholy visit from the blues.

47. "Wenn du bis in die Nächte hinein arbeitest, so handelst du unrecht an dir selber und schlecht an mir. Wozu die Gewalt? Ich habe dir, glaube ich, schon bis zum überdruss wiederholt, dass wir die Mittel haben, (Gottlob, und ich glaube reichlicher, als du dir einbildest!) um dien Studium so lange auszudehnen als wir wollen," Georg Woldemar Cantor to Georg Cantor, January 21, 1863, *Nachlass Cantor I.*
48. Cantor to P.E.B. Jourdain, March 29, 1905, in *Notebook I* of the two Jourdain Notebooks, Archives of the Institut Mittag-Leffler, Djursholm, Sweden. Cantor's letter appears at folio 82. See the transcriptions of the document in GRATTAN-GUINNESS (1971a), 384–385; I. GRATTAN-GUINNESS (1971b). "The Correspondence between Georg Cantor and Philip Jourdain," Jahresber. Dtsch. Math.-Ver. **73:** 123–124.
49. CANTOR, G. (1905): *Ex Oriente Lux. Gespräche eines Meisters mit seinem Schüler über wesentliche Punkte des urkundlichen Christentums. Berichtet vom Schüler selbst Georg Jacob Aaron, cand. sacr. theol. Erstes Gespräch* (Halle, 1905).

    On April 5, 1905, Cantor wrote to Grace Chisholm Young, describing as he had to Jourdain the results of his being "hermetically secluded" in the Halle Nervenklinik for five and a half months:

    "As you know, I had been hermetically secluded 5½ months (from 17. Sept. to 1. March) from the world, except few visits from my family. But I can not say, that I am by this long firebaptism embittered because I do know the great pressure, that has been practiced by the "Ministerium" and the "amiable" german colleagues upon my wife and my children! Further I had a great interest to study the quite unreasonable puerile treatment and soitdisant cure of the lamentable patients. The Muse afforded to me I employed to a renewed study of our Bible with opened eyes and postponing all prejudices. The result has been highly remarkable, as you will see by a little pamphlet (anonymous) of half a sheet, what I will send you perhaps in a week; it is now in the printing office. The title is: 'Ex Oriente Lux. . . .' "

    Cantor wrote the letter in English; it has been printed in GRATTAN-GUINNESS (1971a), 385.
50. The full title referred to essential points "des urkundlichen Christenthums." See note 49.
51. Cantor to Grace Chisholm Young, June 20, 1908, trascribed in H. MESCHKOWSKI (1971): "Zwei unveröffentlichte Briefe Georg Cantors," Der Mathematikunterricht **4:** 30–34.
52. Cantor to Grace Chisholm Young, June 20, 1908. See also W. H. YOUNG (1926): "The Progress of Mathematical Analysis in the Twentieth Century," Proc. London Math. Soc. **24:** 422–423.

    For Poincaré's remarks made in Rome and responsible for Cantor's reply to Mrs. Young, see H. POINCARÉ: "L'avenir des mathématiques," Atti IV Congr. Int. Mat., (Rome, 1909) **1:** 167–182; translated into English as "The Future of Mathematics," Smithsonian Rep. 1909 (Washington: Government Printing Office, 1910), 123–140 (especially 140).
53. CANTOR (1883), 175.
54. Cantor to Klein, February 7, 1883, Letter #432, Archives of the Universitätsbibliothek, Göttingen; Mittag-Leffler to Cantor, March 11, 1883, Archives of the Institut Mittag-Leffler, Djursholm, Sweden.
55. When the *Jahrbuch über die Fortschritte der Mathematik* first began to include the literature of set theory (at Vivanti's behest) in 1894, it was reviewed under the heading "Philosophy"; in 1904 it was given a subsection between "Philosophy" and "Pädagogik"; only after the First World War was set theory given a separate, independent position in the *Jahrbuch.* For further details, see FRAENKEL (1930), 215.
56. Cantor to Heman, June 21, 1888; CANTOR (I), 179.

COMPUTATIONAL COMPLEXITY: ALGORITHMS WITH FEWEST OPERATIONS

Julian D. Laderman

Department of Mathematics
Herbert H. Lehman College
City University of New York
Bronx, New York 10468

Introduction

This paper is concerned with minimizing arithmetic operations. A potential reader might ask oneself "Why should one worry about the number of arithmetic operations when calculators and computers are so fast?" It is true that the mechanics of using a slide rule and the study of logarithms for computational purposes are becoming obsolete. However, the subject area of this paper only came into existence with the advent of high-speed computers and is further enhanced by the development of still faster computers. The reason for this can best be demonstrated by the following problem, which is indicative of this area of study. Consider multiplying two square $n \times n$ matrices A with elements a_{ij} and B with elements b_{jk}. Their product is an $n \times n$ matrix C whose elements are defined by:

$$c_{ik} = \sum_{j=1}^{n} a_{ij} b_{jk}$$

Obtaining the c_{ik} directly from the definition involves n^3 multiplications and $n^2(n - 1)$ additions. The definition is often referred to as the standard or classical algorithm for matrix multiplication. The number of arithmetic operations needed to obtain the product of two square matrices of order n is a function of n. The asymptotic order of this function as n becomes infinite is the asymptotic behavior of the algorithm. For the classical algorithm for matrix multiplication, the asymptotic behavior is $O(n^3)$ (order n^3). Until the late 1960's no results were known on the following two questions:

1) Can it be proved that $O(n^3)$ cannot be reduced?
2) Can algorithms be produced whose asymptotic behavior is less than $O(n^3)$?

The result of efforts to obtain a tight lower bound for the asymptotic order of matrix multiplication is that no bound has been proven higher than $O(n^2)$.

Algorithms, which will be discussed later, have been created which are $O(n^{\log 2 7}) \approx O(n^{2.81})$. This result is the present upper bound for matrix multiplication. Therefore, the two questions mentioned above still remain unanswered if $O(n^3)$ is replaced by $O(n^{\log 2 7})$.

Roughly speaking, this subject area consists of studies involved in trying to answer questions similar to the two mentioned above. Now we can see why such studies are enhanced by the existence of high-speed computers. Consider the size of a problem which is computable by some computing device bound by time. When all computing devices were slow by present standards, it made much less improvement if an

0077–8923/79/0321–0045 $01.75/1 © 1979, NYAS

algorithm of $O(n^2)$ replaced one of $O(n^3)$. This is so because with slow computing devices only a relatively small problem can be handled in either case. These statements are demonstrated by charts in Aho *et al.*[1] (§ 1.1). In addition, it should be realized that as computer hardware improves, the speedup is increased by a constant factor. Therefore, the asymptotic behavior of an algorithm is crucial.

This area of study is referred to by a variety of names. Clearly it is a subarea of "computational complexity," but it would be misleading to call it that. Likewise, "analysis of algorithms" is not totally appropriate. At times, researchers use words like *concrete, arithmetic,* or *algebraic* before the word *complexity* or the phrase *computational complexity*. Since in all problems in this area one is trying to minimize the number of operations, the name *operational complexity* seems most appropriate to the author.

If one wishes to study this area, the only textbook which treats it in depth is that by Aho *et al.*[1] Another relevant source is by Knuth.[2] A very sophisticated, but excellent survey of this area appears in an article by Borodin.[3]

The next section contains the major results on evaluating a polynomial at a point. A shortcut for multiplying two complex numbers can be found in the third section. That result enables one to produce a technique for multiplying two n-digit numbers in fewer multiplications of digits than the usual method. The fourth section is devoted to Winograd's method for obtaining inner products of vectors. A direct result of this is an improved scheme for matrix multiplication. The fifth section discusses the best noncommutative algorithms for matrix multiplication. A technique for creating and analyzing such algorithms appears in the subsequent section. The last section presents the idea of grouping problems into classes of equivalent difficulty.

Polynomial Evaluation

Many researchers have studied the problem of evaluating the polynomial $a_n x^n + a_{n-1}x^{n-1} + \cdots + a_0$ at a point x_0. If one first computes x_0 in successive powers, there are $2n - 1$ multiplications and n additions. A well-known shortcut is Horner's rule where the operations are performed as indicated by the expression:

$$(\cdots((a_n x + a_{n-1})\, x + a_{n-2})\, x + \cdots)\, x + a_0$$

This requires only n multiplications and n additions. The asymptotic improvement over the standard algorithm mentioned above is only by a constant factor. Although this technique is referred to as Horner's rule it was first used by Issac Newton[2] (Vol. 2, p. 423).

Ostrowski[4] proved that at least n multiplications are required for a polynomial of degree $n \leq 4$. Pan[5] has generalized this result to at least n multiplications or divisions for an n-degree polynomial.

To improve polynomial evaluation, Motzkin[6] introduced a technique known as preconditioning where one first obtains some functions of the polynomial's coefficients. These functions can then be employed in evaluating the polynomial at a point x_0. The operations required to obtain these functions are not counted in the total operations to evaluate the polynomial. This might not seem proper, but if the same polynomial is evaluated at many points, it is reasonable since preconditioning must be performed only once. Motzkin has proved that even if preconditioning is used there are at least $n/2$ multiplications or divisions which must occur in evaluating a general n^{th} degree polynomial.

Multiplication of Numbers

An interesting little result deals with the question of how many real multiplications or divisions are required to multiply two complex numbers. Since $(a + ib)(c + id) = (ac - bd) + i(ad + bc)$, four multiplications are usually employed. However, if the three products ac, bd, and $(a + b)(c + d)$ are computed, then $ac - bd$ is immediate and $ad + bc$ is obtained by $(a + b)(c + d) - ac - bd$. Winograd[7] has proved that any algorithm for multiplying two complex numbers requires at least three multiplications or divisions.

The above technique for obtaining the product of two complex numbers with three real multiplications first appeared in Peters.[8] It is similar to a variety of schemes for reducing the number of multiplications of digits to obtain the product of two m-digit binary or decimal numbers. The usual technique, which employs m^2 digit multiplications, was first improved on by Karatsuba and Ofman.[9]

The technique most similar to the one used in multiplying complex numbers will now be described. Consider the product of two decimal numbers x and y each composed of $2n$ digits. Let a represent the left n digits and b the right n digits of x. Similarly, define c and d for y. Now consider obtaining xy from

$$xy = (a10^n + b)(c10^n + d) = ac10^{2n} + (ad + bc)\,10^n + bd$$

The four products of n digit numbers in the last line can be obtained in three multiplications by the technique for complex number multiplication. Of course, there is the extra cost of some shifting and three adding operations. Obvious modifications lead to a procedure for multiplying binary numbers. Aho *et al.* have proved that, using this cutting procedure recursively (with modifications for numbers with odd numbers of digits), the asymptotic result for the total number of digit operations is $O(n^{\log_2 3}) \sim O(n^{1.59})$. This result is not surprising since each time the problem gets cut in half it is replaced by three problems of the reduced size.

Inner Products

In 1968, Winograd[10] created the following algorithm for obtaining inner products. It results in an improved algorithm for computing matrix products. At this time it is necessary to introduce the following notation: [ ] denotes "largest integer less than or equal to."

The inner product of vectors $X = (X_1, X_2, \ldots, X_n)$ and $Y = (Y_1, Y_2, \ldots, Y_n)$ can also be written as

$$\sum_{j=1}^{[n/2]} (X_{2j-1} + Y_{2j})(X_{2j} + Y_{2j-1}) - \alpha - \beta, \quad n \text{ even}$$

and as

$$\sum_{j=1}^{[n/2]} (X_{2j-1} + Y_{2j})(X_{2j} + Y_{2j-1}) - \alpha - \beta + X_n Y_n, \quad n \text{ odd}$$

where

$$\alpha = \sum_{j=1}^{[n/2]} X_{2j-1}X_{2j}; \quad \beta = \sum_{j=1}^{[n/2]} Y_{2j-1}Y_{2j}$$

The following table is obtained when R different inner products are being obtained among N n-dimensional vectors.

	Winograd's Method	Standard Method
Number of Multiplications	$N[n/2] + R[(n+1)/2]$ $= Nn + (R-N)[(n+1)/2]$	$Rn = Nn + (R-N)n$
Number of Additions	$N([n/2]-1)$ $+ R(n + [n/2] + 1)$	$R(n-1)$

Winograd's method requires more additions than the standard method. However, when $R > N$ Winograd's method involves fewer multiplications, and when R is much larger than N, the total number of multiplications is almost cut in half.

Consider the matrix product AB, where A is an $m \times n$ matrix and B is an $n \times p$ matrix. The standard algorithm to obtain AB uses mnp multiplications and $m(n-1)p$ additions. The process actually consists of obtaining mp inner products of $m + p$ n-dimensional vectors. From the above table it is seen that, by Winograd's method, the product of two matrices employs $(m + p)n + (mp - m - p)[(n + 1)/2]$ multiplications and $(m + p)(\lfloor n/2 \rfloor - 1) + mp(n + \lfloor n/2 \rfloor + 1)$ additions. If m, n, and p are increased indefinitely there are asymptotically $\frac{1}{2} mnp$ multiplications and $\frac{3}{2} mnp$ additions.

Matrix Multiplication Algorithms

Strassen[11] created the following algorithm which multiplies two 2×2 matrices in seven multiplications. Assume the matrices are A and B with C as the resulting product. Let

$$m_1 = (a_{11} + a_{22})(b_{11} + b_{22})$$

$$m_2 = (a_{21} + a_{22})b_{11}$$

$$m_3 = a_{11}(b_{12} - b_{22})$$

$$m_4 = a_{22}(-b_{11} + b_{21})$$

$$m_5 = (a_{11} + a_{12})b_{22}$$

$$m_6 = (-a_{11} + a_{21})(b_{11} + b_{12})$$

$$m_7 = (a_{12} - a_{22})(b_{21} + b_{22})$$

then

$$c_{11} = m_1 + m_4 - m_5 + m_7$$

$$c_{12} = m_3 + m_5$$

$$c_{21} = m_2 + m_4$$

$$c_{22} = m_1 + m_3 - m_2 + m_6$$

Strassen does not mention the technique employed to obtain the above famous algorithm. Possibly it was obtained by trial and error with a variety of algebraic identities, such as $ab + cd = a(b - d) + (a + c)d$. A general method for studying the matrix multiplication problem will be discussed in the next section.

In order for this algorithm to be valid, there is no requirement that the matrix elements be commutative under the operation of multiplication. Algorithms with this property can be employed to multiply two matrices whose elements are themselves matrices of appropriate dimensions. These algorithms for matrix multiplication will be referred to as *noncommutative algorithms* and they will be the only ones considered throughout the remainder of this paper. Winograd's algorithm for matrix products in the previous section was not a noncommutative algorithm.

Consider multiplying two matrices of dimension $m^p \times m^p$. This can be looked upon as multiplying two $m \times m$ matrices whose elements are matrices of dimension $m^{p-1} \times m^{p-1}$. Now the matrix products involving matrices of dimension $m^{p-1} \times m^{p-1}$ can in turn be treated as a product of two $m \times m$ matrices whose elements are matrices of size $m^{p-2} \times m^{p-2}$. Therefore it follows that if two $m \times m$ matrices can be multiplied in $f(m)$ multiplications then two $m^p \times m^p$ matrices can be multiplied in $(f(m))^p$ multiplications. If $n = m^p$, then $p = \log_m n$ and $(f(m))^p = f(m)^{\log_m n} = n^{\log_m f(m)}$. Since large matrices can be embedded in matrices of size $n = m^p$ by adding rows and columns of zeros, the asymptotic result for the number of multiplications to multiply two $n \times n$ matrices is $O(n^{\log_m f(m)})$. This asymptotic result is not only valid for the number of multiplications, but it is also appropriate for the total number of arithmetic operations. The number of additions is of the same order as the number of multiplications. Therefore Strassen's algorithm for multiplying two 2×2 matrices yields the asymptotic result $O(n^{\log_2 7}) \approx O(n^{2.81})$ for the total number of operations.

Winograd[7], and Hopcroft and Kerr[12] used two different techniques to prove that seven multiplications must be used in any algorithm to multiply two 2×2 matrices. This result led mathematicians to the study of the multiplication of two 3×3 matrices. Strassen's result immediately led to an algorithm for multiplying two 3×3 matrices in twenty-six multiplications. Gastinel[13] has produced an algorithm using twenty-five multiplications, and Hopcropt and Kerr[12] have produced an algorithm using twenty-four multiplications. Laderman[14] has developed an algorithm that requires only twenty-three multiplications for two 3×3 matrices. Since $O(n^{\log_2 7}) \approx O(n^{\log_3 21.85})$, an algorithm with twenty-one multiplications would be sufficient to improve on Strassen's asymptotic result.

In the product of two 4×4 matrices, Strassen's algorithm used recursively results in an algorithm employing forty-nine multiplications. Therefore an algorithm using forty-eight multiplications would be sufficient to improve on Strassen's result.

In Laderman[15] the 3×3 algorithm with twenty-three multiplications is used in conjunction with other existing algorithms to produce the best (in terms of fewer multiplications) known algorithms for multiplying matrices of many different dimensions. By appropriate partitioning, multiplying two 5×5 matrices can be done in 108 multiplications, two 6×6 matrices can be done in 161 multiplications, two 7×7 matrices can be done in 288 multiplications, and two 9×9 matrices can be done in 529 multiplications. This result for multiplying two 9×9 matrices requires 200 fewer multiplications than the classical algorithm. Actually it is shown that there is an improvement in an infinite number of cases. One need only consider multiplying two square matrices of order 3×2^m by partitioning each matrix into nine square blocks of order 2^m.

Development of 3 × 3 Algorithm

The technique that produced the algorithm for multiplying two 3×3 matrices in twenty-three multiplications will be outlined in this section. The problem of searching for algorithms for a matrix product with t multiplications will be converted to the problem of finding an integer solution to a large system of equations. Suppose we have a noncommutative algorithm involving t multiplications for $C = AB$, where A is an $m \times n$ matrix and B is $n \times p$. Then there exists t linear forms $f_1, f_2, \ldots, f_t$ which depend only on the elements of A, and an additional t linear forms $g_1, g_2, \ldots, g_t$ which depend only on the elements of B, such that c_{ik} can be expressed as a linear combination of their products.

That is

$$c_{ik} = \sum_{v=1}^{t} Z_{ikv} f_v g_v,$$

where

$$f_v = \sum_{j,h} x_{jhv} a_{jhv}; \quad 1 \le j \le m,\ 1 \le h \le n$$

and

$$g_v = \sum_{r,s} y_{rsv} b_{rsv}; \quad 1 \le r \le n,\ 1 \le s \le p$$

Combining the above lines one gets

$$c_{ik} = \sum_{v,j,h,r,s} z_{ikv} x_{jhv} a_{jhv} y_{rsv} b_{rsv}$$

From the definition of matrix multiplication we have the following constraints for the appropriate system of equations. Considering an element c_{ik} we need the sum of the products to be 0 except in the case where $i = g$, $h = r$, and $s = k$. In this case the sum of the products should be 1. Therefore,

$$\sum_{v=1}^{t} z_{ikv} x_{jhv} y_{rsv} = \begin{cases} 1, \text{ if } i = g,\ h = r,\ s = k \\ 0, \text{ otherwise} \end{cases}$$

There are $m^2n^2p^2$ equations and $(mp + mn + np)t$ unknowns. Brent[16] had previously searched for a real solution of these equations with the use of a computer. A real solution could improve the asymptotic result for matrix multiplication, but it would not be practical to use for obtaining matrix products.

When we consider finding an algorithm to multiply two 3×3 matrices in twenty-three multiplications, the system of 729 equations involving 621 unknowns is as follows:

$$\sum_{v=1}^{23} z_{ikv}x_{jhv}y_{rsv} = \begin{cases} 1, \text{ if } i = g,\ h = r,\ s = k \\ 0, \text{ otherwise} \end{cases}$$

where $i, k, j, h, r, s = 1, 2,$ or 3.

If we consider only solutions of the equations where the variables take on values from the set $\{-1, 0, 1\}$, the products of three variables can only take on values from this same set. This enables one to display a solution of the equations in a convenient way. Then by a technique demonstrated by Laderman[15] it is possible to study solutions of similar, but smaller systems of equations and then merge four such systems to yield a solution from which the twenty-three multiplication algorithm follows.

The technique for solving systems of equations of this type may produce an algorithm to multiply two 3×3 matrices in twenty-one or fewer multiplications or two 4×4 matrices in forty-eight or fewer multiplications. Either of these solutions would yield an improvement on Strassen's asymptotic result.

Problems Equivalent to Matrix Multiplication

Another study which falls into this subject area deals with analyzing two problems (called A,B) in order to demonstrate that if A has asymptotic order K, then B has asymptotic order K; likewise if B has asymptotic order M, then A has asymptotic order M. Using this relationship equivalence classes of problems have been developed.

As shown by Aho *et al.*[1] (Chapter 6), in the class of matrix multiplication, there are the important problems of inverting a matrix, evaluating a determinant, and solving a system of n linear equations in n unknowns. These equivalent problems further enhance the interest in developing a matrix multiplication algorithm with lower asymptotic order.

Two special cases of matrix multiplication will now be considered. The problem of squaring an arbitrary matrix and proving it is in the same equivalence class as the matrix multiplication problem will be treated first. Obviously, if two matrices could be multiplied in order K, then a matrix could be squared in order K. Now consider the following identity

$$\begin{pmatrix} 0 & A \\ B & 0 \end{pmatrix}^2 = \begin{pmatrix} AB & 0 \\ 0 & BA \end{pmatrix}$$

where A and B are both $n \times n$ matrices.

Therefore, the problem of squaring matrices of dimension $2n$ requires at least as many operations as multiplying two $n \times n$ matrices. Therefore, only the coefficients of their asymptotic behavior can differ and they are in the same class. This simple proof appears in Munro[17], where it is also proved that the problem of multiplying two symmetric matrices is of the same asymptotic order as the general matrix multiplication problem. Obviously an algorithm for the general matrix multiplication problem will work for the symmetric matrices so half the proof is completed. Now consider the following identity

$$\begin{pmatrix} 0 & A \\ A^T & 0 \end{pmatrix} \begin{pmatrix} 0 & B^T \\ B & 0 \end{pmatrix} = \begin{pmatrix} AB & 0 \\ 0 & A^T B^T \end{pmatrix}$$

where A and B are both $n \times n$ matrices and A^{T} denotes the transpose of A.

Clearly multiplying two symmetric matrices each of dimension $2n$ requires at least as many operations as multiplying two $n \times n$ matrices. Therefore, they are in the same equivalence class.

Summary

This paper is concerned with computing one or more mathematical expressions with a minimal number of total arithmetic operations or with a minimal number of multiplications and divisions. We are concerned with questions of proving that a computation requires at least a specific number of operations and with the problem of creating new algorithms that employ fewer operations than any known algorithm. Particular emphasis is on matrix multiplication and problems of equivalent computational difficulty.

References

1. Aho, A. V., J. E. Hopcroft & J. D. Ullman. 1974. The Design and Analysis of Computer Algorithms. Addison-Wesley, Reading, Mass.
2. Knuth, D. E. The Art of Computer Programming. 1968 Vol. 1, 1969 Vol. 2, 1973 Vol. 3. Addison-Wesley, Reading, Mass.
3. Borodin, A. B. 1973. On the number of arithmetics required to compute certain functions. *In* Complexity of Sequential and Parallel Numerical Algorithms. J. F. Traub, Ed. Academic Press, New York, N.Y. pp. 149–180.
4. Ostrowski, A. M. 1954. On two problems in abstract algebra connected with Horner's rule. (Studies presented to R. von Mises.) Academic Press, New York, N.Y., pp. 40–48.
5. Pan, V. Y. 1966. Methods of computing values of polynomials. Russ. Math. Surv. **21**: 105–136.
6. Motzkin, T. S. 1955. Evaluation of polynomials and evaluation of rational functions. Bull. Am. Math. Soc. **61**: 163.
7. Winograd, S. 1971. On multiplication of 2 × 2 matrices. Linear Alg. Appl. **4**: 381–388.
8. Peters, T. R. 1963. Evaluation of polynomials with the minimum number of operations. New York University. M. S. thesis.
9. Karatsuba, A. & Y. Ofman. 1962. Multiplication of multidigit numbers on automata. Dokl. Akad. Nauk. SSSR. **145**: 293–294.
10. Winograd, S. 1968. A new algorithm for inner product. IEEE Trans. Comput. **17**: 693–694.
11. Strassen, V. 1969. Gaussian elimination is not optimal. Numer. Math. **13**: 354–356.

12. HOPCROFT, J. E. & L. R. KERR. 1971. On minimizing the number of multiplications necessary for matrix multiplication. SIAM J. Appl. Math. **20**: 30–36.
13. GASTINEL, N. 1971. Sur le calcul des produits de matrices. Numer. Math. **17**: 222–229.
14. LADERMAN, J. D. 1976. A noncommutative algorithm for multiplying 3×3 matrices using 23 multiplications. Bull. Am. Math. Soc. **82**: 126–128.
15. LADERMAN, J. D. 1976. On algorithms for minimizing the number of multiplications in matrix products. New York University. Doctoral dissertation.
16. BRENT, R. P. 1970. Algorithms for matrix multiplication. Tech. Rep. STAN-CS-70-157. Department of Computer Science, Stanford University.
17. MUNRO, I. 1971. Problems related to matrix multiplication. Proc. Courant Inst. Symp. Comput. Complexity. pp. 137–151.

CRUDE THEORETICAL MODELS OF JET-DRIVEN WIND INSTRUMENTS AND THE EDGETONE

Harry E. Rauch

Graduate School
City University of New York
New York, New York 10036

Introduction

If one makes the request, "please tell me how the trumpet or clarinet works, but do it quickly, simply, and without a lot of mathematics or other theory," then a ready response is found in the classic work of Lord Rayleigh[1] (§322k). If a little more detail is desired at the price of a little mathematics then §46 of the same work suffices. Rayleigh attributes his response to Helmholtz,[2] (Appendix VIII), but Benade[3] cites Wilhelm Weber, 1830, as an earlier source.

A similar request for enlightenment about the operation of the flute or flue organ pipe (or even a blown bottle) brings a simple, but unfortunately long since discredited, response, again from Rayleigh (ibid.) and attributed to Helmholtz (ibid., pp. 92–93), and again, the writer surmises from a citation by Coltman,[4] anticipated in 1830, this time by Sir William Herschel. The discredited theory, or model, will be sketched in the next section.

If one pursues the matter in search of a valid model, one is soon lost in a swamp of technical papers and controversy. Authors on all sides express grave reservations about attempts at simplified theories. One early investigator[5] proposed a simplified theory of the edgetone, which proved inaccurate.

However, the excellent work, theoretical and experimental, of several modern investigators, some of them motivated by the study of the closely related phenomenon of the edgetone, has revealed several basically simple mechanisms which, this writer feels, can be assembled into crudely simplified models of the desired type—models which, according to one's taste, are tolerably consistent with reality, and plausible or a good target for critics to shoot at. The investigators whose works were drawn upon by the writer are Powell,[11] Cremer and Ising,[6] Coltman,[4] and, later, for the edgetone, Karamcheti and sometime coworkers, Shields[8] and Stegen.[9]

The resulting models like those of the reed-controlled instruments are small-signal, linear ones. A tone generator is *linear* if it can produce a pure (sinusoidal) tone. No real generator is linear, but, if tone amplitudes are sufficiently small, most real generators are sufficiently linear. In other words, the present models, if they have any validity at all, have more validity for softly playing instruments or softly sounding edgetones

The writer takes this opportunity to call attention to some cogent observations of one of the referees. The first is that the present paper deals only with the feedback mechanism of the flute, but not with its resonator, particular tone quality, or the effect of shape, design, and material on the tone quality. The model here is a very limited one in this sense. The second observation is that the proposed criteria of plausibility of the model (the last section) may be difficult to verify. He proposes that it might be

0077–8923/79/0321–0054 $01.75/1 © 1979, NYAS

possible to realize a digital version of the model to be compared with known digital models of real flutes.

In hindsight, it is remarkable that the basic mechanism for the new models was sketched by Rayleigh[1] (p. 412) in one paragraph (the last of §371) on the birdcall, a cylindrical version of the (planar) edgetone.

Jet-Driven Instruments—The Old, Discredited Model

Briefly the (Herschel ?)-Helmholtz-Rayleigh model of, say, a flute is that of *instantaneous action:* when the air flows *into* the mouthhole in unison with the vibrating air column in the flute body, the jet is sucked into the flute and reinforces the flow in the air column, thereby compensating energy losses due to friction, etc.; when the air flows *out* of the mouthhole, again in unison with the air column, the jet is pushed out of the flute and has no effect.

This model cannot be a correct one. Leaving aside the somewhat subtle question of whether the mouthhole is a point of zero pressure (it is not), one finds that the crucial objection has long been evident, namely, that the flute or any other jet-driven instrument only sounds when the jet velocity and mouth-to-edge or orifice-to-edge distance satisfy certain conditions. This is clear for the flute, and it had long been noted, for example, how critical is the "cut-up" on a flue organ pipe. In the instantaneous action model these quantities play no role, and the model must be rejected as being unsatisfactory.

A Modern Model of the Jet-Driven Instruments—A Sketch

The way to a model of jet-driven instruments which accounts for dependence on jet (blowing) velocity and orifice to edge distance was found via studies of a related phenomenon, the *edgetone,* which will be discussed in a later section. This work suggests that the underlying mechanism is the propagation of jet disturbances from the orifice, where they originate, to the edge, where they cause the jet to enter or leave the air column. In particular, it suggests that the frequency of the sounded tone depends on the time the jet disturbance takes to travel from orifice to edge. Since that time, it develops, depends on the jet velocity and, manifestly, on the orifice-edge distance, one may hope that a model based on the jet disturbance mechanism will have some chance of accounting for the observed critical dependence of the operation of the instrument on these quantities. The model must also account, however, for the *appearance* of instantaneous action, which has misled earlier investigators and is confirmed by more recent observations, namely, that the jet in *normal* operation (see below) does move into and out of the air column in unison with the motion of the air in the column.

For the benefit of the reader who wishes to avoid any unnecessary technical discussion, the writer will now immediately formulate the model, reserving a more technical description for later. For convenience of visualization let the instrument be the flute. The orifice is the outside of the opening between the player's lips. The edge is the far edge of the mouthhole.

The jet and the air in the column are assumed to move in and out of the mouthhole together, switching from *in* to *out* and *vice versa* instantaneously. The course of one cycle, i.e., one acoustical vibration, is described as follows. The air moves out of the mouthhole (and the jet moves out) and "disturbs" the jet at the orifice. This "out" disturbance travels toward the edge and, *by hypothesis, causes* the jet to move *in,* i.e., into the mouthhole. But by the first assumption this is only possible if the air column is already drawing the jet *in* at the same instant. That is, the time required for the disturbance initiated at the orifice (by the *out* motion of the air column through the mouthhole) to reach the edge and "kick" the jet *in* must be the same as the time required for a *half-cycle.* The other half-cycle is described mutatis mutandis. That is the model. It solves the problem of apparent instantaneous action, and it solves the problem of dependence on jet velocity and orifice-edge distance.

A Modern Model of the Jet-Driven Instruments—More Precision

There remains the question of the validity, i.e., reality of the model of this section. To postpone controversy as long as possible it seems advisable, at the price of introducing another hypothesis, to deduce some quantitative properties of the model. If these agree with reality, the (strengthened) model will at least be *consistent* with reality no matter how implausible the individual hypotheses. If the individual hypotheses can be shown to approximate reality, then the model becomes plausible. Some definitions are useful.

A *jet-edge system* is a thin, flat jet of air issuing from an orifice and impinging on an edge lying in the plane of the jet and normal to its direction of flow.

A *resonator* is an enclosed body of air whose container has an aperture called the *mouth* (there may be other apertures).

In the spirit of the "black box" one defines an *ideal* jet-edge system to be one satisfying the following hypotheses when oscillating (as part of a sounding instrument or edgetone):

(i) *The jet output hypothesis.* The jet (at the edge, understood henceforth) is in one of two states, labeled *out* and *in,* with the indicated meaning, when the jet-edge system is at the mouth of a resonator, or *up* and *down* when the edgetone is under consideration, in which case the plane of the jet is pictured as horizontal. The transition between the two states is (virtually) instantaneous.

(ii) *The jet input hypothesis.* There are two inputs to the jet at the orifice: *out* and *in* mouth volume-displacements (with a resonator) or *up* and *down* transverse displacements (in the edgetone case) of the air into which the jet issues at the orifice.

(iii) *The jet signal hypothesis.* Either input at the orifice causes a signal to be transmitted along the jet from orifice to edge with a nonzero time delay. When the signal arrives at the edge the jet switches, i.e., changes its state.

(iv) *The phase reversal hypothesis.* The input *out* (up) at the orifice results in the output *in* (down) at the edge and *vice versa.*

(v) *The hypothesis of constant signal velocity.* This is a strengthening of part of (iii). It demands that the signal be propagated with a velocity that is a constant fraction c of the jet velocity u, where c is also independent of position along the jet and, *for a certain band of frequencies,* the frequency of oscillation. Coltman[4] gives a typical value of $c = .4$.

An ideal jet-edge system coupled to a resonator at the mouth will be called an *ideal flute.*

An additional hypothesis for an ideal flute is useful:

(vi) *The hypothesis of normal operation* or *apparent instantaneous action.* The jet output states *out* and *in* occur simultaneously with the outward and inward motions of the mouth volume-displacement of the resonator, respectively.

An ideal flute satisfying (vi) is said to be in *normal operation.*

Ideal flutes, like all abstractions, do not exist in reality. Real flutes are not even linear, let alone, do they satisfy any one of the first five hypotheses. The sixth hypothesis has a somewhat different character. One can conceive of an ideal flute *not* in normal operation but with some fixed phase difference between jet in (out) and mouth-volume displacement in (out) due to dissipative effects just as in real flutes. The point of all this is that the "kick" of the jet from *out* to *in* or *vice versa* as a result of the signal from the orifice is the decisive control element, *not* the direct push or pull of the jet by the displacement of the air through the mouth, as intuition might lead one to expect.

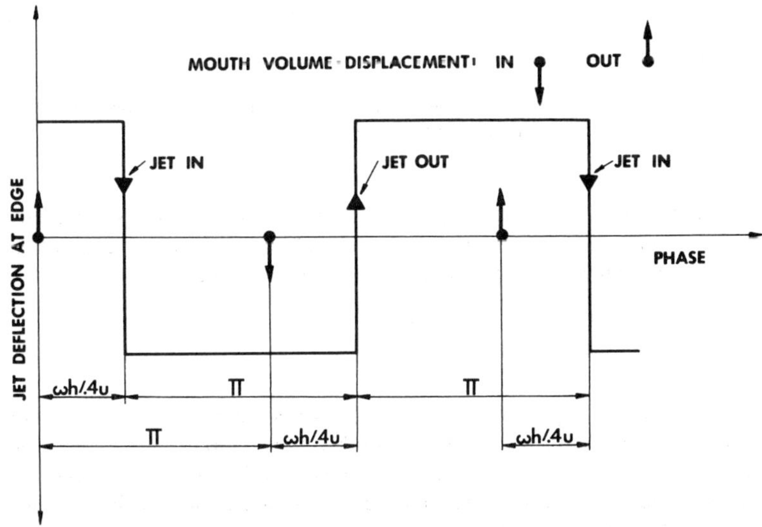

FIGURE 1. Jet deflection at edge versus phase for jet-driven instruments assuming only hypotheses (i)–(v), and that jet does not switch in the interval $\omega h/.4u$ after mouth volume-displacement switches. Positive deflection on the graph means jet out.

Now some quantitative properties of an ideal flute. If, as before, u is the jet velocity and cu the jet disturbance velocity (i.e., the signal velocity), let h be the orifice-edge distance. Then, by (v), the transit time, or *time lag,* for the signal is

$$\tau = \frac{h}{cu} \tag{1}$$

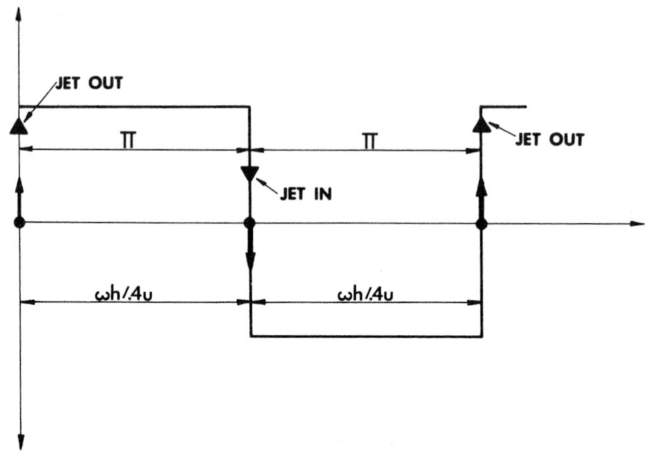

FIGURE 2. Jet deflection at edge versus phase (instruments) for normal operation and assumptions of FIGURE 1. Conventions are the same as in FIGURE 1.

Assume now that the flute is sounding a pure tone, i.e., oscillating in simple harmonic motion, with frequency f and angular frequency $\omega = 2\pi f$ (here is the assumption of linearity). The *phase lag* of the jet due to the time lag in Equation 1 is

$$\omega\tau = \frac{\omega h}{cu} = \frac{2\pi f}{cu} \tag{2}$$

Consider the situation depicted in FIGURE 1. It is assumed there that at phase (and time) zero, the jet is already *out* and the mouth volume-displacement is *starting out*. By (iii), (iv), and (v), at phase $\omega h/cu$ the jet will switch. It is assumed further that the jet has not switched at all in that interval. Then, at phase $\omega h/cu + \pi$ the jet will switch *out*. Before that, however, at phase π the mouth volume-displacement will go *in* and will have *caused* the last mentioned switch *out* by the jet. And so on, the mouth volume-displacement will oscillate at the same frequency as the jet, but will lead the jet by phase $\omega h/cu$. There is *no restriction* on the frequency, or, to put it another way, none on h and u.

Now assume the ideal flute is in *normal operation,* i.e., assume (vi), too. Then clearly FIGURE 1 becomes FIGURE 2, and one sees from the figure that

$$\frac{\omega h}{cu} = \pi, \tag{3}$$

or

$$f = \frac{cu}{2h}. \tag{4}$$

To put it in words, Equation 3, and thus Equation 4, is a consequence of the fact that, if the jet and the mouth volume-displacement go *in* together, then the *in* signal for the jet to switch *in* must have been sent exactly π earlier, phasewise. But π earlier the mouth volume-displacement is *out*. Therefore, the *in* signal must have resulted from an *out* displacement at the orifice of the jet. One sees then that the *phase reversal hypothesis* (iv) is essential to compensate for the fact that the jet deflection at the edge lags the jet input (mouth volume-displacement) at the orifice by π, if the jet at the edge and the mouth volume-displacement are to be in phase. This deserves the name "jet paradox," if one considers that, without phase reversal, the only way that jet switching and mouth volume-displacement switching could occur *simultaneously* is if they were in *antiphase*. Consistency of Equation 4 with experiment and plausibility of the hypotheses will be examined in subsequent sections.

The Edgetone—An Analogous Model

A *real* jet-edge system, *not* coupled to a resonator, can, for certain values of jet velocity u and orifice-edge distance h, sound a tone, called the *edgetone*. The mechanism replacing the resonator feedback seems to be a transverse displacement of the air at the orifice induced (at the speed of sound, i.e., virtually instantaneously) by the jet at the edge as it switches state.

Accordingly, if it be assumed that an ideal jet-edge system sound a pure tone (simple harmonic oscillation), it will be called an *ideal edgetone*. Here it must be understood that the locutions *up, down,* and *transverse displacement* are to be used in the hypotheses (i)–(v) in this section.

For reasons developed in a subsequent section, the substitution for (vi) is:

(vi, *bis*) *The hypothesis of normal operation for the edgetone.* The transverse displacement at the orifice *leads* the jet deflection at the edge by the phase $\pi/2$.

A glance at FIGURE 3, where the case $k = 1$ is illustrated, will confirm that

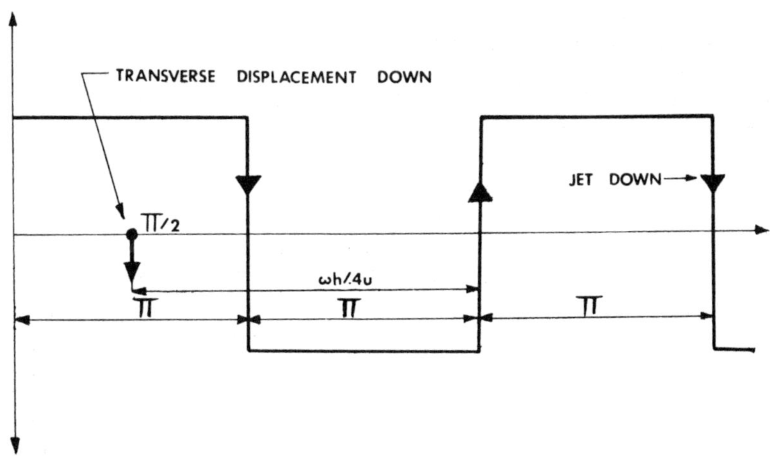

FIGURE 3. Jet deflection at edge versus phase for normal operation of edgetone at stage 1(k = 1). Conventions are the same as in FIGURE 1.

$$\frac{\omega h}{cu} = 2k\pi - \frac{\pi}{2} \tag{5}$$

where k is a positive integer called the *stage,* is consistent with the action of the *ideal edgetone in normal* operation. But Equation 5 yields

$$f = \left(k - \frac{1}{4}\right)\frac{cu}{h} \tag{6}$$

The quantity $k - 1/4$ in Equation 6 is the *phase fraction* (at the edge). Denoting it by α, one can think of it as that fraction of a whole wave of jet disturbance lying between the orifice and the edge, and one can write a more general formula,

$$f = \frac{\alpha cu}{h} \tag{7}$$

for the frequency of an ideal edgetone, where for normal operation in stage 1 $\alpha = 0.75$. It should be noted that

$$\alpha c = \frac{fh}{u} = S, \tag{8}$$

where S is the *Strouhal number* (dimensionless frequency) based on orifice-edge distance.

Consistency of Both Models with Reality

Equations 4 and 6 or 7 or 8 will be used as criteria for consistency of the respective models with reality. It would be ideal if the experimental data known to the writer contained, besides the frequency f, or, equivalently, the Strouhal number S, the values of c for jet-driven instruments and *both* c and α for edgetones. Unfortunately, for the former this is not always the case and for the latter it is never the case. One set of data (Brown's, see below) gives neither α nor c. Accordingly, the writer has resorted to a number of schemes to supply values of α or c, or both, to make the models fit the data

Table 1

Jet-Driven Instruments

Instrument	c	u cm/sec	h cm	f exp Hz	$f = cu/2h$ Hz
Flute[4]	.41	1460	.7	440	428
Flue organ pipe[6]	.4	1800	3	123	120
	.41	1800	3	123	—
Flue organ pipe[10]	.5	700	.7	250	250
Flue organ pipe[10]	.5	1300	.7	495	464
Flue organ pipe[10]	.5	1500	.7	510	538

TABLE 2

Stage	f Hz	$S = fh/u$	α	$c = S/\alpha$
1	370	.374	.90	.416
2	815	.879	1.81	.486
1 during 2	3.6	.319	.92	.35

as closely as possible. One suggested measure of consistency then becomes the mutual compatibility of the various resulting α (or c).

In TABLE 1 this situation occurs once, namely for the organ pipe discussed in Cremer and Ising.[6] Despite their FIGURE 8 they do not give an *experimental* value of c for their pipe operating at $f = 123$ Hz and $u = 1800$ cm/sec. It is dealt with the two ways. In the first line the typical value $c = .4$ of Coltman[4] is inserted in Equation 4 to compute f. In the second line the experimental value of f is inserted in Equation 4 to compute c. In the subsequent lines of TABLE 1 it should be noted that the values of c were computed by means of outside excitation of the jet by a sinusoidal source and not from the instruments themselves while they were in self-excited oscillation. The maximum deviation of 6% of computed f from experimental f is encouraging.

The experimental data for the edgetone are less satisfactory. One reason may be, first of all, the indications that some of the experiments reported were well into the nonlinear range of operation where hypotheses (v) and (vi, *bis*) are in question. Secondly, accurate measurement of α or c is difficult.

Brown gives the empirical formula[7]

$$f = .466j(u - 40)\left(\frac{1}{h} - .07\right), j = 1, 2.3, 3.8, 5.4, \tag{9}$$

for his experimental edgetone frequencies in stages 1, 2, 3, 4, respectively, where u is in cm/sec and h is in cm, but no values of *either* c or α. On comparison of Equation 9 with Equation 6, $k = 1, 2, 3, 4$, one finds from the latter the frequency ratios 1, 2.3, 3.7, 5, which are reasonably consistent with the values of j in the former. But if one assumes $\alpha = .75$ in Equation 7 for stage 1 and compares Equation 7 with Equation 9, one finds, on ignoring the subtrahends, $c = .466/.75 = .62$, which seems inconsistently high compared to the values in TABLE 1. However, if one assumes $\alpha = .9$, as reported by Stegen and Karamcheti (see TABLE 2), then $c = .52$, which is better.

Shields and Karamcheti[8] (p. 24) report $S = .365$ ($f = 419$ Hz) and (p. 45) $\alpha = .58$, so that in the model $c = .365/.58 = .63$, which seems irreconcilable with the values of TABLE 1. Even though $\alpha = .58$ is tolerably consistent with $\alpha = .75$ it seems that no amount of playing with α and c can make the model applicable to this particular edgetone.

TABLE 2 lists experimental data from Stegen and Karamcheti[9] together with experimental Strouhal members and values of c computed on the basis of Equation 8. The stage 2 phase fraction is in good agreement with 1.75 obtained from Equation 6 with $k = 2$. The phase fraction for stage 1 is not so good, but probably nonlinearity, in particular, nonapplicability of the model, holds. Even so, the computed values of c seem reasonable.

TABLE 3, based on Coltman,[10] is analogous to TABLE 2, but lists the c given there and the α computed from it. The values of c are subject to the same criticism as those from the same reference in TABLE 1. Again the computed values of α are tolerably close to .75.

PLAUSIBILITY OF THE MODEL

The hypotheses of the two previous sections were abstracted from Coltman,[4] Cremer and Ising,[6] and Powell.[11] The Coltman[10] and Fletcher[12] papers were unknown to the writer at the time he formulated the models, and, indeed the Coltman paper postdates it, while the Fletcher paper seems to offer a different model for the jet-driven instruments from that advocated here.

Hypotheses (i)–(iii) seem explicit or implicit in all the quoted sources. The writer suspected the phase reversal hypothesis and then found it in Powell,[11] who gives arguments, some based on experiment, of its plausibility. Subsequently, the writer realized that Cremer and Ising implicitly used it (see their Equation 35, which is equivalent to phase reversal in their model). Hypothesis (v) was suggested to the writer by reading Coltman.[4] The correctness of such hypotheses has been vigorously attacked by Karamcheti and coworkers on the basis of their experimental work, and, strictly speaking, they are right. Nonetheless, even their work, and certainly Coltman's, tends to show a certain crude plausibility of (v) within certain ranges of experimental parameters—witness TABLES 2 and 3. Hypothesis (v) has also been attacked because it is inconsistent with classical linear stability theory for jets (Rayleigh's equation). But that theory is a homogeneous eigenvalue theory, and what is involved here is an inhomogeneous forcing theory, which is not adequately developed to the writer's knowledge.

TABLE 3

Stage	f Hz	S	c	$\alpha = S/c$
1	175	.306	.5	.61
1	220	.308	.5	.62
1	315	.315	.5	.63
1	470	.329	.5	.66
1	620	.334	.5	.67
1	720	.336	.5	.67

The plausibility of (vi) is self-evident. Not so, hypothesis (vi, *bis*). Indeed, the writer's first version used the word *lags* instead of *leads* and was suggested by Powell's arguments. A careful reading of his arguments suggests the phase difference $\pi/2$ with an ambiguity of sign. Powell opted for *lag,* but subsequent experimental work by Karamcheti and coworkers has shown that assumption inconsistent with reality and they also questioned the $\pi/2$. On becoming aware of their work, the writer changed *lag* to *lead.*

This concludes the writer's sources and the arguments which he believes make the hypotheses, and hence the models, plausible.

SUMMARY

There exists a "bare-bones" small-signal, linear theory, going back at least to Helmholtz and Rayleigh, of reed-controlled wind instruments, such as the trumpet and clarinet. Here, an attempt is made to formulate similar (i.e., crudely oversimplified) models for the jet-driven instruments, such as the flute, recorder, and organ flue pipes on the basis of the theoretical and experimental contributions of several modern investigators, some of whom were motivated by the study of the *edgetone*. The edgetone is the sound obtained when a thin, flat jet of air hits the edge of a wedge or a thin wire held in the plane of the jet at right angles to the stream when jet velocity and the distance from jet orifice to edge satisfy certain conditions. An attempt is also made to model the edgetone.

ACKNOWLEDGMENTS

The writer wishes to thank his colleague, Richard Sacksteder, for calling his attention to the mathematics and physics of wind instruments and supplying basic references, as well as for helpful conversations.

REFERENCES

1. STRUTT, J. W. S. (Lord Rayleigh). 1945. Theory of Sound. Dover, New York.
2. HELMHOLTZ, H. 1954. On the Sensations of Tone. Dover, New York.
3. BENADE, A. H. 1973. Scientific American (July). **229:** 24–35.
4. COLTMAN, J. M. 1968. J. Acoust. Soc. Am. **44:** 983–992.
5. KÖNIG, W. 1912. Phys. Z. **13:** 1053–1054.
6. CREMER, L. & H. ISING. 1967. Acustica. **19:** 143–153.
7. WOOD, A. 1966. Acoustics. Dover, New York.
8. SHIELDS, W. L. & K. KARAMCHETI. 1967. SUDAAR No. 304, Stanford.
9. STEGEN, G. R. & K. KARAMCHETI. 1970. J. Sound Vib. **12:** 281–284.
10. COLTMAN, J. M. 1976. J. Acoust. Soc. Am. **60:** 725–733.
11. POWELL, A. 1961. J. Acoust. Soc. Am. **33:** 395–409.
12. FLETCHER, N. H. 1974. J. Acoust. Soc. Am. **56:** 645–652.

STRUCTURE AND STABILITY IN WEIGHTED DIGRAPH MODELS*

Fred S. Roberts

Department of Mathematics
Rutgers University
New Brunswick, New Jersey 08903

Complex Systems and Structural Modeling

Many of the problems our society faces are concerned with large, complex, and difficult-to-understand systems. This is true of problems involving energy supply and demand, food supply, transportation, pollution, urban services, health care, and so on. The application of mathematical techniques, so frequently successful in dealing with small, well-defined, and mechanistic physical or engineering problems, is not always very helpful in dealing with these more complex systems. However, mathematics has a role to play, if we are more modest in what we expect to learn about a system whose behavior is being studied mathematically. In this paper, we shall describe some techniques which are part of a general approach to the study of complex systems known as structural modeling. In structural modeling, we concern ourselves mostly with qualitative predictions about systems. We are mostly interested in learning about "geometric" properties of these systems, properties which describe their type of growth, their stability or instability, their sensitivity, and so on, in qualitative terms: they grow without bound, they oscillate wildly, etc. More specifically, we are interested in relating these qualitative properties of a system to certain structural properties of the system.

Techniques of structural modeling have been applied to such diverse problems as energy use, air pollution, and transportation systems,[12,22–27,29] health care delivery, water policy, and environmental policy,[7–10] naval manpower,[11] the analysis of coastal resources and the transportation of coal in inland waterways,[1,5] and the study of ecosystems.[14,15] Recent surveys of structural modeling have been carried out by Cearlock,[2] and Lendaris and Wakeland.[13]

We shall concentrate on the technique of pulse process analysis developed by Roberts,[22–27] and Roberts and Brown,[29] and generalized by McLean[16] and McLean and Shepherd.[17–19] We shall solve some of the problems posed in Roberts[24] and extend some of the results of Roberts and Brown.[29] We shall discuss applications of pulse process analysis to problems of health care delivery, energy use in food production, and transportation.

Weighted Digraphs

The pulse process analysis starts with a weighted digraph representing basic relations among variables underlying a complex system. To be precise, a *digraph* $D = (V, A)$ consists of a finite set V of *vertices* and a subset A of $V \times V$, called the set of

*Supported by NSF Grant Number MCS76–06336.

0077–8923/79/0321–0064 $01.75/1 © 1979, NYAS

arcs. We shall allow *loops,* arcs of the form (x,x). A *weighted digraph* consists of a digraph together with a *weight* $w(x,y)$, a nonzero real number, on each arc (x,y). The existence of an arc (x,y) will be interpreted to mean that, all other things being equal, a change in x has a significant direct effect on y. The weight will be interpreted in general terms as the strength of the effect. We shall introduce a more specific interpretation in the next section. Both positive and negative numbers are allowed as weights. We interpret a positive weight on arc (x,y) to mean that whatever happens at x results in a change "in the same direction" in y, while a negative weight means that whatever happens at x results in a change "in the opposite direction" in y.

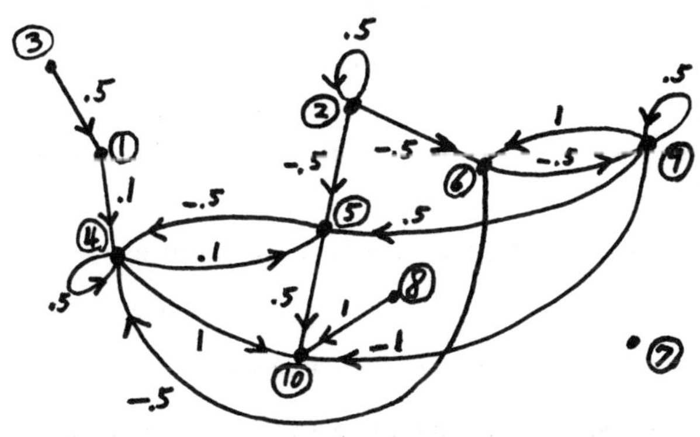

Key

1. Environmental stresses
2. Age structure
3. Population size
4. Morbidity rate
5. Screening efforts (searching out health problems)
6. Availability of health care services
7. Physician time available
8. Health care facilities
9. Delegation of physician duties to other personnel
10. Expenditures

FIGURE 1. Weighted digraph for health care delivery in British Columbia (from data in Kane *et al.*[8]).

FIGURES 1, 2, and 3 show weighted digraphs. FIGURE 1, based on data from Kane,[8] was constructed to study health care delivery in British Columbia. FIGURE 2, from Roberts,[28] was constructed to study the increasing use of energy in food production; it was built in the context of the observation[20,21] that, while yields of a field devoted to a corp, such as corn, increased dramatically between 1945 and 1970, by a factor of about 2.5, the energy input required to produce the increased yield increased by an even larger factor, a factor of about 3. FIGURE 3, based on data from Kane,[7] was constructed to study the pattern of public and private transportation in Vancouver.

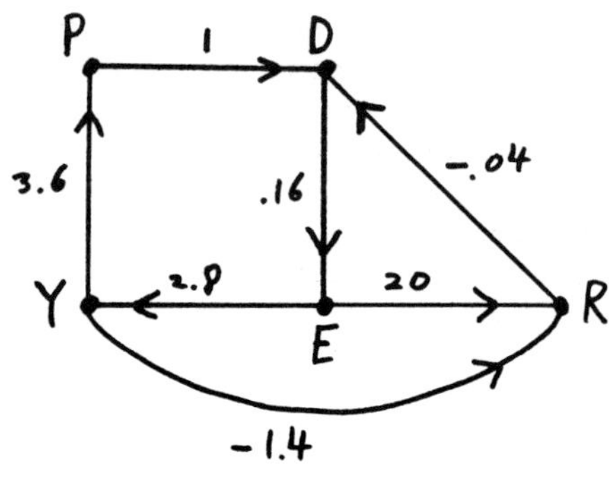

Key

P Population
D Demand for food
Y Food yield
R Cost of food
E Energy input into food production

FIGURE 2. Weighted digraph for the food-energy system.[28]

PULSE PROCESSES

The weights in a weighted digraph will have for us a very specific interpretation. Namely, we shall assume that every variable has a level or *value,* which changes over time and is measured at discrete times, such as each day, week, year, etc. We shall speak of a time period, and mean a day, week, year, or whatever unit of time is appropriate. Then:

RULE P. If there is an arc from x to y with weight $w(x,y)$, an increase in variable x by u units leads to an increase in variable y by $u \times w(x,y)$ units one time period later.

For example, in FIGURE 2, the weight 2.8 on the arc (E,Y) means that whenever energy input into food production goes up by u energy units, then food yield goes up by $2.8u$ food yield units. The weight -1.4 on the arc (Y,R) means that an increase in food yield of u food yield units leads to a decrease in the cost of food by $1.4u$ cost of food units. The weights in FIGURES 1 and 3 were in fact determined under somewhat different interpretations. However, for the sake of illustration of the ideas, we shall interpret the weights there in this way as well.

The interpretation of weights given in RULE P implies some very special assumptions about the way changes take place:

(1) The effect of a change in x on y is linear in the amount of the change in x.
(2) The effect of a change in x on y is independent of the levels of x and y and the time at which the change takes place.
(3) Each effect represented by an arc takes the same amount of time.

RULE P is said to define an *autonomous pulse process* on a weighted digraph provided we give as initial conditions the values of all variables at the start (time 0) and the changes induced in these variables at the start. To be specific, suppose $u_1, u_2, \ldots, u_n$ are the variables (vertices of the weighted digraph), $v_j(t)$ is the value at u_j at time t and $p_j(t)$ is the change or *pulse* at u_j at time t. The pulse $p_j(t)$ is defined to be $v_j(t) - v_j(t-1)$ provided $t > 0$, and it is given as an initial condition provided $t = 0$. Suppose A is the *adjacency matrix,* the n × n matrix whose i,j entry is $w(u_i,u_j)$ if there is an arc from u_i to u_j and 0 if there is no such arc. Let

$$V(t) = (v_1(t), v_2(t), \ldots, v_n(t))$$

and

$$P(t) = (p_1(t), p_2(t), \ldots, p_n(t))$$

Then the autonomous pulse process model can be summarized as follows:

$$V(t + 1) = V(t) + P(t + 1),$$
$$P(t + 1) = P(t)A.$$

Stability under Pulse Processes

A vertex u_j is said to be *pulse stable* under an autonomous pulse process if

$$\{\,|p_j(t)|: t = 0, 1, \ldots\}$$

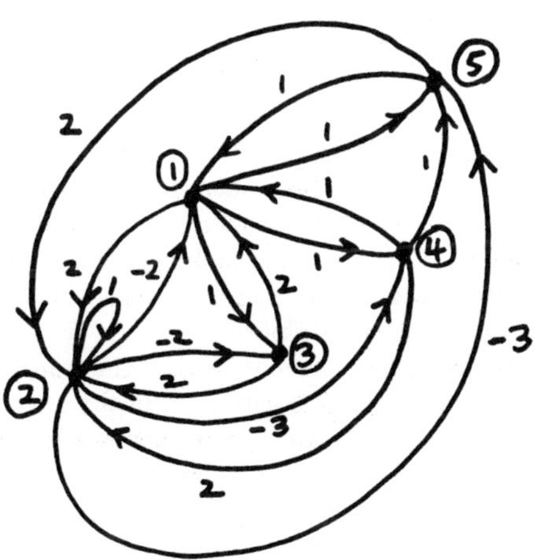

Key

1. Cost of an automobile
2. Use of automobiles
3. Comfort and convenience of automobile use
4. Freedom of choice in travel (routes, time, etc.)
5. Speed

FIGURE 3. Weighted digraph for pattern of public/private transportation in Vancouver (from data in Kane[7]).

is bounded, and *value stable* under the autonomous pulse process if

$$\{\,|v_j(t)|: t = 0, 1, \ldots\}$$

is bounded. A weighted digraph is *pulse* or *value stable* under an autonomous pulse process if each vertex is, respectively, pulse or value stable. It is easy to see that pulse and value stability are independent of initial *values* and depend only on initial pulses, that value stability implies pulse stability, that an exponentially growing variable is unstable in both senses, and that a linearly growing variable is value unstable but pulse stable.[26,29]

The first theorems about pulse and value stability which we shall state deal with the eigenvalues of the adjacency matrix A. Let J be the Jordan Canonical Form corresponding to A. We say an eigenvalue λ of A is *linked* if in J, the submatrix,

$$\begin{pmatrix} \lambda & 1 \\ 0 & \lambda \end{pmatrix}$$

appears. Of course, only multiple eigenvalues can be linked.

THEOREM 1. A weighted digraph is pulse stable under all autonomous pulse processes if and only if every eigenvalue of its adjacency matrix has magnitude less than or equal to 1 and every linked eigenvalue has magnitude less than 1.[29]

THEOREM 2. A weighted digraph is value stable under all autonomous pulse processes if and only if it is pulse stable under all autonomous pulse processes and 1 is not an eigenvalue of its adjacency matrix.[29]

Before illustrating these theorems, let us comment on what they say. To say a weighted digraph D is pulse unstable under some autonomous pulse process means that there is some initial set of pulses or changes which lead to unbounded pulses or changes at some variable. Thus, the pulse stability conclusion says that no matter how initial changes are introduced, no variable has unboundedly large changes over time. Similar comments apply to value stability.

To illustrate these theorems, let us begin with the weighted digraph of FIGURE 1. Here, the eigenvalues turn out to be 0 with multiplicity 5, and .5, .14, .36, $.25 + .66i$, and $.25 - .66i$. Since these eigenvalues all have magnitude less than 1, THEOREM 1 implies that the digraph is pulse stable under all autonomous pulse processes and, since 1 is not an eigenvalue, THEOREM 2 implies the digraph is value stable under all autonomous pulse processes. If we were to observe, for example, continually growing (health care) expenditures—i.e., if, empirically, $v_j(t)$ seemed to be growing without bound for the 10th variable—then we could not explain this observation on the basis of the internal workings of the system. If any initial stresses or pulses in the system are left to work themselves out *internally* according to the rules of autonomous pulse processes, they would not lead to increasing values at any variable. We could try to modify the pulse process model to try to get a prediction of value instability. Alternatively, we could salvage the model by simply pointing out that there are other ways for a variable to increase without bound, for example, by getting repeated *externally* generated pulses. We would then have to look for an explanation of

increasing health care expenditures by studying variables beyond those in the system modeled by the weighted digraph of FIGURE 1.

In the weighted digraph of FIGURE 2, the characteristic polynomial of the adjacency matrix turns out to be

$$C(\lambda) = \lambda(\lambda^4 + .128\lambda - 1.638) \tag{1}$$

Since the product of the roots of a polynomial with highest coefficient 1 is plus or minus the constant term, we see that 1.638 is plus or minus the product of the nonzero eigenvalues, and, in particular, it follows that there must be an eigenvalue of magnitude larger than 1. Hence, the weighted digraph is pulse unstable under some autonomous pulse process, and therefore also value unstable. This is not surprising, in light of the observations of rapidly growing population, food prices, and energy use in food production. In contrast to the health care example, this growth could be explained on the basis of the internal workings of the system.

Finally, in the weighted digraph of FIGURE 3, the characteristic polynomial turns out to be

$$C(\lambda) = \lambda^5 - \lambda^4 + 16\lambda^3 + 41\lambda^2 + 7\lambda - 2 \tag{2}$$

Again, the product of the roots is $\pm$ 2, and so there is a nonzero eigenvalue of magnitude larger than 1. We conclude that the digraph is pulse and value unstable under all autonomous pulse processes.

STRUCTURE AND STABILITY

One trouble with theorems like THEOREMS 1 and 2 is that they do not pinpoint the source of an instability such as that predicted for the digraph of FIGURE 2. In particular, these theorems do not help us to determine possible changes of the digraph (the system being modelled by the digraph) which might lead to stability. As we previously pointed out in Roberts[24] and Roberts and Brown,[29] one would like to discover theorems which relate stability results, or more specifically eigenvalues, to structural properties of the digraph. The next few sections state such results.

STRONG COMPONENTS

A *path* in a digraph (V, A) is a sequence $u_1, u_2, \ldots, u_k$ of vertices such that, for all $i = 1, 2, \ldots, k\text{-}1$, (u_i, u_{i+1}) is an arc. We say (V, A) is *strongly connected* if, for all u, v in V, there is a path from u to v and a path from v to u. A *subgraph* of (V, A) is a digraph (W, B) such that W is a subset of V and B is a subset of $A \cap (W \times W)$. The subgraph (W, B) is *generated* if $B = A \cap (W \times W)$, i.e., if B consists of all arcs of A joining two vertices of W. Finally, a *strong component* of (V, A) is a maximal, strongly connected, generated subgraph. It is easy to see that the strong components partition the vertices of a digraph. Generated subgraphs and strong components of weighted digraphs D are just generated subgraphs of the underlying digraph using the weights of D as weights.

Our first structure theorem is the following:

THEOREM 3. The characteristic polynomial of (the adjacency matrix of) a weighted digraph is the product of the characteristic polynomials of (the adjacency matrices of) its strong components.[6]

COROLLARY. The set of eigenvalues of a weighted digraph is the union (counting multiplicity) of the set of eigenvalues of its strong components.

To illustrate THEOREM 3 and its corollary, we note that the weighted digraph of FIGURE 1 has the following strong components: {1}, {2}, {3}, {7}, {8}, {10}, {4, 5}, and {6, 9}. Hence, instead of calculating the characteristic polynomial of a 10 × 10 matrix, we simply have to calculate characteristic polynomials of six 1 × 1 and two 2 × 2 matrices. We obtain

$$C(\lambda) = \lambda\lambda\lambda\lambda\lambda(\lambda - .5)(\lambda^2 - .5\lambda + .05)(\lambda^2 - .5\lambda + .5) \tag{3}$$

To compute eigenvalues, we note that all of the singleton strong components except {2} have eigenvalue 0, while {2} has eigenvalue .5. The strong component {4, 5} has eigenvalues .14 and .36, while the strong component {6, 9} has eigenvalues .25 ± .66i, giving us the ten eigenvalues we stated above. THEOREM 3 and its COROLLARY are not very helpful with the weighted digraphs of FIGURES 2 and 3, since these digraphs are strongly connected and so there is only one strong component.

1-FACTORS

A path $u_1, u_2, \ldots, u_k$ in a digraph is called a *cycle* if $u_1, u_2, \ldots, u_{k-1}$ are distinct and $u_k = u_1$. A digraph is a *1-factor* if every vertex has exactly one incoming and one outgoing arc. Every digraph consisting of a single cycle is a 1-factor. More generally, it is easy to see that a 1-factor is a union of disjoint cycles.

A *spanning subgraph* of a digraph (V, A) is a subgraph (W, B) in which $W = V$. A *1-factor of a digraph* is defined to be a spanning subgraph which is a 1-factor. For example, in the digraph D of FIGURE 3, one 1-factor F is shown in FIGURE 4. Notice that the 1-factor F is a proper spanning subgraph of D, since arcs such as (1,4) are not used.

If F is a 1-factor in a weighted digraph D, the *weight* $w(F)$ is defined to be the

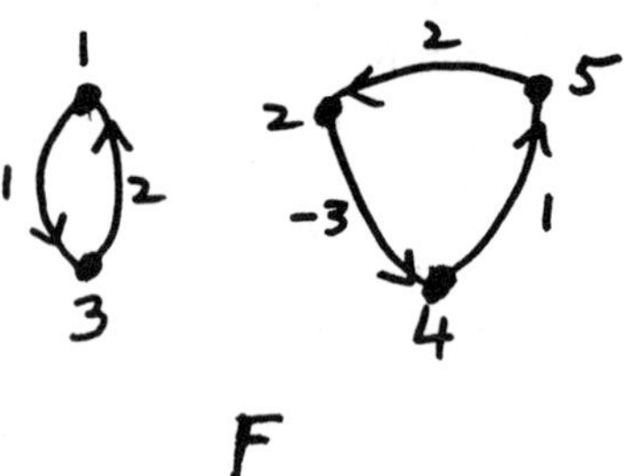

FIGURE 4. A 1-factor F in the weighted digraph of FIGURE 3.

TABLE 1
COMPUTATION OF $C(\lambda)$ FOR THE WEIGHTED DIGRAPH OF FIGURE 1

k	Generators	1-factor	Weight	Parity	b_k
1	2	2,2	.5	1	$b_1 = (-1)(.5 + .5 + .5) = -1.5$
	4	4,4	.5	1	
	9	9,9	.5	1	
2	2,4	2,2 + 4,4	.25	2	$b_2 = (1)(.25 + .25 + .25)$
	2,9	2,2 + 9,9	.25	2	$+ (-1)(-.05 - .5) = 1.3$
	4,9	4,4 + 9,9	.25	2	
	4,5	4,5,4	$-.05$	1	
	6,9	6,9,6	$-.5$	1	
3	2,4,5	4,5,4 + 2,2	$-.025$	2	$b_3 = (1)(-.025 - .25 - .025$
	2,4,9	2,2 + 4,4 + 9,9	.125	3	$- .25) + (-1)(.125)$
	2,6,9	6,9,6 + 2,2	$-.25$	2	$= -.675$
	4,5,9	4,5,4 + 9,9	$-.025$	2	
	4,6,9	6,9,6 + 4,4	$-.25$	2	
4	2,4,5,9	4,5,4 + 2,2 + 9,9	$-.0125$	3	$b_4 = (1)(.025)$
	2,4,6,9	6,9,6 + 2,2 + 4,4	$-.125$	3	$+ (-1)(-.0125 - .125)$
	4,5,6,9	4,5,4 + 6,9,6	.025	2	$= .1625$
5	2,4,5,6,9	4,5,4 + 6,9,6 + 2,2	.0125	3	$b_5 = (-1)(.0125) = -.0125$
6	none				$b_6 = 0$
7	none				$b_7 = 0$
8	none				$b_8 = 0$
9	none				$b_9 = 0$

product of the weights of the arcs of F and the *parity* p(F) is the number of cycles in F. Thus, in our previous example, w(F) is -12, while p(F) is 2. Given a digraph D and an integer k, let

$$b_k = \sum_{E_k}\sum_{u}(-1)^{p(F_u)}w(F_u) \tag{4}$$

where the first sum is over all generated subgraphs E_k of D of k vertices and the second sum is over all 1-factors F_u of E_k. For example, in the weighted digraph of FIGURE 1, if $k = 2$, there are only five generated subgraphs of two vertices which have 1-factors. These are the subgraphs generated by the following sets of vertices: $\{2,4\}$, $\{2,9\}$, $\{4,9\}$, $\{4,5\}$, and $\{6,9\}$. In each case, there is just one 1-factor in each generated subgraph. These 1-factors have weights .25, .25, .25, $-.05$, and $-.5$, respectively. The first three 1-factors have parity 2, and the last two have parity 1. Hence,

$$b_2 = 1(.25) + 1(.25) + 1(.25) + (-1)(-.05) + (-1)(-.5) = 1.3$$

A similar computation for the other values b_k is shown in TABLE 1.

THEOREM 4. If D is a weighted digraph and $C(\lambda)$ is the characteristic polynomial of D (of D's adjacency matrix), then

$$C(\lambda) = \lambda^n + \sum_{k=1}^{n} b_k \lambda^{n-k} \tag{5}$$

where b_k is given by (4).[3,4]

To illustrate this theorem, we observe that according to TABLE 1, for the weighted digraph of FIGURE 1,

$$\begin{aligned} C(\lambda) &= \lambda^{10} - 1.5\lambda^9 + 1.3\lambda^8 - .675\lambda^7 + .1625\lambda^6 - .0125\lambda^5 \\ &= \lambda^5(\lambda^5 - 1.5\lambda^4 + 1.3\lambda^3 - .675\lambda^2 + .1625\lambda - .0125). \end{aligned}$$

TABLE 2

COMPUTATION OF $C(\lambda)$ FOR THE WEIGHTED DIGRAPH OF FIGURE 3.

k	Generators	1-factor	Weight	Parity	b_k
1	2	2,2	1	1	$b_1 = (-1)(1) = -1$
2	1,2	1,2,1	−4	1	$b_2 = (-1)(-4 + 2 + 1 + 1$
	1,3	1,3,1	2	1	$- 4 - 6 - 6) = 16$
	1,4	1,4,1	1	1	
	1,5	1,5,1	1	1	
	2,3	2,3,2	−4	1	
	2,4	2,4,2	−6	1	
	2,5	2,5,2	−6	1	
3	1,2,3	1,2,3,1	−8	1	$b_3 = (1)(2 + 1 + 1)$
	1,2,3	1,3,2,1	−4	1	$+ (-1)(-8 - 4 - 6 - 4 - 6$
	1,2,3	1,3,1 + 2,2	2	2	$- 4 + 1 - 6) = 41$
	1,2,4	1,2,4,1	−6	1	
	1,2,4	1,4,2,1	−4	1	
	1,2,4	1,4,1 + 2,2	1	2	
	1,2,5	1,2,5,1	−6	1	
	1,2,5	1,5,2,1	−4	1	
	1,2,5	1,5,1 + 2,2	1	2	
	1,4,5	1,4,5,1	1	1	
	2,4,5	2,4,5,2	−6	1	
4	1,2,4,5	1,2,4,5,1	−6	1	$b_4 = (1)(-6 - 6 + 1 - 4 - 12$
	1,2,4,5	2,1,4,5,2	−4	1	$- 4 - 12) + (-1)(-6 - 4$
	1,2,4,5	1,4,1 + 2,5,2	−6	2	$- 6 - 8 - 6 - 8) = 7$
	1,2,4,5	2,4,2 + 1,5,1	−6	2	
	1,2,4,5	1,4,5,1 + 2,2	1	2	
	1,2,3,4	1,3,2,4,1	−6	1	
	1,2,3,4	1,4,2,3,1	−8	1	
	1,2,3,4	2,3,2 + 1,4,1	−4	2	
	1,2,3,4	1,3,1 + 2,4,2	−12	2	
	1,2,3,5	1,3,2,5,1	−6	1	
	1,2,3,5	1,5,2,3,1	−8	1	
	1,2,3,5	2,3,2 + 1,5,1	−4	2	
	1,2,3,5	1,3,1 + 2,5,2	−12	2	
5	1,2,3,4,5	1,3,1 + 2,4,5,2	−12	2	$b_5 = (1)(-12 - 4)$
	1,2,3,4,5	2,3,2 + 1,4,5,1	−4	2	$+ (-1)(-6 - 8) = -2$
	1,2,3,4,5	1,3,2,4,5,1	−6	1	
	1,2,3,4,5	2,3,1,4,5,1	−8	1	

The reader can check that $C(\lambda)$ agrees with the formula given in (**3**).

As a second illustration of the theorem, let us consider the weighted digraph of FIGURE 2. Since every pair of cycles has a common vertex, the only 1-factors in generated subgraphs are the three cycles of the digraph: F_1 = D,E,R,D; F_2 = P,D,E,Y,P; F_3 = D,E,Y,R,D. Since $p(F_1) = p(F_2) = p(F_3) = 1$ and $w(F_1) = -.128$, $w(F_2) = 1.613$, and $w(F_3) = .025$, we can see that $b_1 = b_2 = b_5 = 0$, $b_3 = (-1)(-.128) = .128$, and $b_4 = (-1)(1.613) + (-1)(.025) = -1.638$. It follows that $C(\lambda)$ is given by (**1**).

As a final illustration of the theorem, we observe that, according to TABLE 2, for the weighted digraph of FIGURE 3,

$$C(\lambda) = \lambda^5 - \lambda^4 + 16\lambda^3 + 41\lambda^2 + 7\lambda - 2$$

as we stated previously in (**2**).

In general, THEOREM 4 can be applied as follows. Suppose $C(\lambda)$ factors as $\lambda^{n-s}(\lambda^s + \ldots + b_s)$, $b_s \neq 0$. We know that the product of the nonzero eigenvalues is $\pm b_s$, and hence pulse stability implies that $|b_s| \leqq 1$. In the example of FIGURE 2, we have already observed that $|b_s| > 1$. We could conceivably attain stability by pushing $|b_s|$ down. One way to do this in our example is to break up a cycle with a high weight, such as P,D,E,Y,P. In particular, deleting the arc (Y,P) will accomplish this purpose. Deletion of this arc results in a weighted digraph with the following characteristic polynomial:

$$C(\lambda) = \lambda(\lambda^4 + .128\lambda - .025)$$

A simple calculation shows that the eigenvalues are now all of magnitude less than 1 and hence we have pulse and value stability under all autonomous pulse processes. The deletion of the arc (Y,P) corresponds to some form of population control: population is not allowed to increase with increasing availability of food.

INTEGER WEIGHTS

A weighted digraph is called *integer-weighted* if every weight is an integer (positive or negative). FIGURE 5 gives an example of an integer weighted digraph.

THEOREM 5. Suppose D is an integer-weighted digraph. Then D is pulse stable under all autonomous pulse processes if and only if every nonzero eigenvalue of its adjacency matrix has magnitude equal to 1 and no nonzero eigenvalue is linked.[29]

FIGURE 5. An integer-weighted digraph.

This theorem says that if we happen to know that the weights are integers, then in the pulse stable case, there can be no nonzero eigenvalues of magnitude less than 1. The next two theorems, whose proofs we shall comment on below, state conditions for stability in an integer-weighted digraph which can be related to the structure of the digraph.†

THEOREM 6. Suppose D is an integer-weighted digraph and its characteristic polynomial is

$$C(\lambda) = \lambda^n + \sum_{k=1}^{n} b_k \lambda^{n-k}$$

Suppose some $b_k \neq 0$ and s is the highest subscript such that $b_s \neq 0$. If D is pulse stable under all autonomous pulse processes, then

$$b_s = \pm 1 \tag{6}$$

and

$$b_i = b_s b_{s-i},\ i = 1, 2, \ldots, s-1 \tag{7}$$

THEOREM 7. Under the hypotheses of THEOREM 6, D is value stable under all autonomous pulse processes if and only if D is pulse stable under all autonomous pulse processes and

$$\sum_{k=1}^{s} b_k \neq -1. \tag{8}$$

THEOREMS 6 and 7 can be combined with the result of THEOREM 4 to relate structure to stability. To illustrate, in the integer-weighted digraph of FIGURE 3, according to TABLE 2, $s = 5$ and $b_s = -2$. Thus, by THEOREM 6, D is pulse unstable under some autonomous pulse process. Hence, by THEOREM 2 or 7, D is also value unstable under some autonomous pulse process. If D is the integer-weighted digraph of FIGURE 5, then it is easy to see that $s = 2$, $b_1 = -1$, and $b_2 = -1$. Then condition (**6**) holds but condition (**7**) fails for $i = 1$. It follows that D is pulse unstable under some autonomous pulse process, and hence that D is also value unstable under some autonomous pulse process. Suppose the loop from b to b is removed from the digraph D. In the resulting digraph, $s = 2$, $b_1 = 0$, and $b_2 = -1$. Thus, conditions (**6**) and (**7**) are satisfied, which gives us no conclusion. However, $C(\lambda) = \lambda^2 - 1$, and so, by THEOREM 1 or 5, this digraph is pulse stable under all autonomous pulse processes. However, condition (**8**) is violated, so the digraph is value unstable under some autonomous pulse process. (This is also clear by THEOREM 2, since 1 is an eigenvalue.)

THEOREMS 6 and 7 generalize some results of Roberts and Brown.[29] We say a digraph D is an *advanced rosette* if D is strongly connected and there is a vertex v

†In practice, we rarely know weights exactly. This means that we have to do a sensitivity analysis, checking stability or instability conclusions for sensitivity under small change in weights. It follows that, even if the weights in the particular digraph we obtain are integers, in doing a sensitivity analysis we will have to check a noninteger-weighted case. One would like to have results similar to the following for the arbitrary weighted case.

which is on every cycle of D. Suppose D is an advanced rosette with weights $+1$ or -1 on each arc. The *weight* of a cycle is the product of the weights of its arcs. Let a_k be the sum of the weights of the cycles of length k, i.e., the number of cycles of length k with an even number of -1 weights less the number of cycles of length k with an odd number of -1 weights. Assuming that some $a_k \neq 0$, let s be the highest subscript such that $a_s \neq 0$. Then, since every pair of cycles in an advanced rosette has a common vertex v, it follows, from THEOREM 4, that $a_k = -b_k$, where b_k is given in (**4**). Roberts and Brown prove the following: Suppose D is an advanced rosette with weights $+1$ or -1 on each arc, and suppose

$$C(\lambda) = \lambda^{n-s} (\lambda^s - \sum_{k=1}^{s} a_k \lambda^{s-k}) \qquad \textbf{(9)}$$

Then if D is pulse stable under all autonomous pulse processes, we have

$$a_s = \pm 1 \qquad \textbf{(10)}$$

and

$$a_i = -a_s a_{s-i},\ i = 1, 2, \ldots, s-1. \qquad \textbf{(11)}$$

Moreover, D is value stable under all autonomous pulse processes if and only if D is pulse stable under all autonomous pulse processes and

$$\sum_{k=1}^{s} a_k \neq 1 \qquad \textbf{(12)}$$

An analysis of the proof shows all that is needed are integer weights and the fact that $C(\lambda)$ is given by (**9**). Now under the hypothesis $b_k = 0$, for $k > s$, (**9**), (**10**), (**11**), and (**12**) readily translate into (**5**), (**6**), (**7**), and (**8**). Hence, the proof given by Roberts and Brown really proves THEOREMS 6 and 7.‡

SUMMARY

Weighted digraphs are described as models of complex systems. Pulse processes on weighted digraphs and notions of stability of a weighted digraph under a pulse process are defined. Results are stated which relate stability to the properties of the cycles and, more generally, of the 1-factors of a weighted digraph. These results generalize the results of Roberts and Brown on stability in advanced rosettes. They are applied to problems of health care delivery, energy use in the production of food, and transportation.

ACKNOWLEDGMENT

The author is grateful for the research assistance of Robert Opsut.

‡The author thanks Professor Philip Straffin for making essentially this observation and the observation that the b_k could be obtained using a result like Chen's.

REFERENCES

1. ANTLE, L. G. & G. P. JOHNSON. 1973. Integration of policy simulation, decision analysis, and information systems: implications of energy conservation and fuel substitution measures on inland waterway traffic. Proceedings of Computer Science and Statistics Seventh Annual Symposium on the Interface. Iowa State University, Ames, Iowa.
2. CEARLOCK, D. B. 1977. Common properties and limitations of some structural modeling techniques. University of Washington, Seattle, Washington. Doctoral Dissertation.
3. CHEN, W. K. 1967. On directed graph solutions of linear algebraic equations. SIAM Rev. **9:** 692–707.
4. CHEN, W. K. 1971. Applied Graph Theory. American Elsevier, New York, N.Y.
5. COADY, S. K., G. P. JOHNSON & J. M. JOHNSON. 1973. Effectively conveying results: a key to the usefulness of technology assessment. Mimeographed, Institute of Water Resources, Corps of Engineers. Presented at First International Congress on Technology Assessment (May). The Hague, the Netherlands.
6. HARARY, F. 1959. A graph-theoretic method for the complete reduction of a matrix with a view toward finding its eigenvalues. J. Math. Phys. **38:** 104–111.
7. KANE, J. 1972. A primer for a new cross-impact language–KSIM. Technol. Forecasting Soc. Change. **4:** 129–142.
8. KANE, J., W. THOMPSON & I. VERTINSKY. 1972. Health care delivery: a policy simulation. Socio-Econ. Plan. Sci. **6:** 283–293.
9. KANE, J., I. VERTINSKY & W. THOMPSON. 1972. Environmental simulation and policy formulation—methodology and example (water policy for British Columbia). A. K. Biswas, Ed., International Symposium on Modeling Techniques in Water Resources Systems. Environment. Ottawa, Canada.
10. KANE, J., I. VERTINSKY & W. THOMPSON. 1973. KSIM: A methodology for interactive resource policy simulation. Water Resources Res. **9:** 65–79.
11. KRUZIC, P. G. 1973. A suggested paradigm for policy planning. Stanford Res. Inst. Tech. Note TN-OED-016 (June).
12. KRUZIC, P. G. 1973. Cross-impact analysis workshop. Stanford Res. Inst. Letter Rep. (June).
13. LENDARIS, G. G. & W. W. WAKELAND. 1977. Structural modeling—a bird's eye view. Syst. Sci. Ph.D. Program. Portland State University, Portland, Oregon (February).
14. LEVINS, R. 1974. Problems of signed digraphs in ecological theory. S. Levin, Ed. Ecosystem Analysis and Prediction. SIAM Publications, Philadelphia, Pa. pp. 264–277.
15. LEVINS, R. 1974. The qualitative analysis of partially specified systems. Ann. N. Y. Acad. Sci. **231:** 123–138.
16. MCLEAN, M. 1976. Getting the problem right—The role for structural modeling. Science Policy Research Unit. University of Sussex, Sussex, England. Mimeographed.
17. MCLEAN, M. & P. SHEPHERD. 1976. The importance of model structure. Futures. **8:** 40–51.
18. MCLEAN, M. & P. SHEPHERD. 1978. Feedback processes in dynamic models. Technol. Forecasting Soc. Change. **11:** 153–164.
19. MCLEAN, M. & P. SHEPHERD. The United Kingdom private car model—An example of structural modeling. Technol. Forecasting Soc. Change. Submitted for publication.
20. PIMENTEL, D., L. E. HURD, A. C. BELLOTTI, M. J. FORSTER, I. N. OKA, O. D. SHOLES & R. J. WHITMAN. 1973. Food production and the energy crisis. Science. **182:** 443–449.
21. PIMENTEL, D., W. R. LYNN, W. K. MACREYNOLDS, M. T. HEWES & S. RUSH. 1974. Workshop on research methodologies for studies of energy, food, man and environment, Phase 1. Tech. Rep. (June). Cornell University Center for Environmental Quality Management, Ithaca, N.Y.
22. ROBERTS, F. S. 1971. Signed digraphs and the growing demand for energy. Environ. Plan. **3:** 395–410.
23. ROBERTS, F. S. 1973. Building and analyzing an energy demand signed digraph. Environ. Plan. **5:** 199–221.
24. ROBERTS, F. S. 1974. Structural characterizations of stability of signed digraphs under pulse processes. R. Bari & F. Harary, Eds. Graphs and Combinatorics. Lect. Not. #406. Springer-Verlag, Berlin-Heidelberg-New York. pp. 330–338.

25. Roberts, F. S. 1975. Weighted digraph models for the assessment of energy use and air pollution in transportation systems. Environ. Plan. **7:** 703–724.
26. Roberts, F. S. 1976. Discrete Mathematical Models, with Applications to Social, Biological, and Environmental Problems. Prentice-Hall, Englewood Cliffs, N.J.
27. Roberts, F. S. 1976. Structural analysis of energy systems. F. S. Roberts, Ed. Energy: Mathematics and Models. SIAM Publications, Philadelphia, Pa. pp. 84–101.
28. Roberts, F. S. 1978. Graph Theory and its Applications to Problems of Society. NSF-CBMS Monograph 29. SIAM Publications, Philadelphia, Pa.
29. Roberts, F. S. & T. A. Brown. 1975. Signed digraphs and the energy crisis. Am. Math. Mon. **82:** 577–594.

VIBRATIONS IN FLUID DYNAMICS

Richard Sacksteder

Department of Mathematics
Graduate Center
City University of New York
New York, New York 10036

Eighteenth Century Fluid Dynamics

One of the basic problems of fluid dynamics is to describe what happens when a fluid flows around a solid body. Here the discussion will be limited to the case where the body is the interior of a simple closed curve Γ situated in the plane and the flow takes place in the exterior domain E of Γ. Euler's method of describing the kinematics of such a flow is to specify vectors $V(y, t)$ denoting the velocity of the particle of fluid that is at the point y of E at the time t. The vector field V must satisfy the condition

$$\lim V(y, t) = V_\infty, \quad |y| \to \infty \qquad \textbf{(1.1)}$$

This corresponds to the fact that far away from the body the fluid is expected to approach an overall stream velocity. The assumption that the body is really solid is expressed by:

$$\text{If } y \text{ is on } \Gamma,\ V(y, t) \text{ is tangent to } \Gamma. \qquad \textbf{(1.2)}$$

It sometimes makes physical sense to require in addition that

$$\int_E |V(y, t) - V_\infty|^2\, dy < \infty \qquad \textbf{(1.3)}$$

This means that the change in energy of the fluid brought about by putting the body into the stream is finite; however, in many applications (**1.3**) is not assumed. (Here, dy is the usual element of area in the plane.)

It is perhaps fortunate that Euler and the other early investigators were in no position to know the fine points about the properties of fluids. In fact, Euler's famous equations for fluid flow are based on only two assumptions, conservation of matter and conservation of momentum. If one assumes, as we shall here, that the fluid is incompressible, conservation of matter means that the area filled up by any "part" of the fluid remains constant. This condition is expressed in terms of V by

$$\operatorname{div} V = \nabla \cdot V = 0 \qquad \textbf{(1.4)}$$

The acceleration of the particle that is at y at time t is

$$\frac{DV}{Dt}(y, t) = V_t(y, t) + \nabla_V V(y, t)$$

0077–8923/79/0321–0078 $01.75/1 © 1979, NYAS

where the subscript denotes partial differentiation and ∇_V is covariant differentiation. Then the momentum equation is

$$\frac{DV}{Dt} = \nabla Q \tag{1.5}$$

where Q is a function of y and t that can be interpreted in terms of pressure and density.

It is fairly easy to find a vector field V satisfying the conditions (**1.1**)–(**1.5**), as well as

$$\nabla \times V = \text{curl } V = 0, \; V \text{ independent of } t \tag{1.6}$$

Such solutions are essentially unique, but if (**1.3**) is not required, there is a whole one-parameter family of solutions satisfying (**1.1**), (**1.2**), (**1.4**), and (**1.5**), as well as (**1.6**). These solutions in some ways reflect properties that are expected and they are much used in applications, but they have serious deficiencies. The most famous of these is D'Alembert's paradox[2], that is, the failure to predict drag forces on the body; but there is another that seems worthy of receiving serious attention from a modern point of view. This is that no solution requiring that V be independent of t is in accord with common facts of everyday experience, such as the whistling of wind in the trees and the flapping of flags in the breeze. It is such problems that will concern us here.

Experiments

Strouhal[1,6,10] performed the first careful experiments aimed at measuring the frequency of the vibrations that are generated when a fluid flows past a solid body. His apparatus consisted of a long rod that could be rotated with a known angular velocity around an axis parallel to the direction of the rod. He determined the frequency of the vibrations produced from the pitch of the faint sounds that were heard. The results were, approximately, that if d denotes the diameter of the rod, U the velocity of its center, and f the frequency of the sounds emitted, then $f = .185U/d$, and f is independent of everything else. It is reasonable to identify the results obtained with what would be expected in a planar flow around a disk of diameter d with U identified with V_∞. Accordingly, the dimensionless number $S = fd/V_\infty$, called the Strouhal number, was estimated to be given by

$$S = .185 \tag{2.1}$$

Rayleigh[10] took an interest in Strouhal's experiments and made two important contributions. First, he showed experimentally that there were periodically varying forces acting on the rod (or disk) and that, contrary to what many might expect, they acted perpendicularly, rather than parallel, to the flow direction V_∞. Subsequently, Rayleigh analyzed Strouhal's data further and performed some experiments of his own using water in place of air. He argued that, on dimensional grounds, S should be given by a formula of the type

$$S = a(1 - bR^{-1}) \tag{2.2}$$

where R^{-1} (the reciprocal of the Reynolds number) is a dimensionless quantity that, everything else being equal, is proportional to the viscosity of the fluid. On the basis of the data available he suggested that $a = .195$ and $b = 20.1$. Later experiments have led to the suggestions $a = .198$, $b = 19.7$ and $a = .21$, $b = 20$.[1,6] Of course, it is not expected that (**2.2**) is valid for all values of R, but only in some range such as $50 \leq R \leq 100{,}000$, because, for high values of R, there are turbulent effects and for low values Rayleigh's arguments, which were based on expansion in powers of R^{-1}, break down. However, (**2.2**) is reasonally valid for a range that covers many practical applications.

The attempts to calculate S theoretically have not met with impressive success. An account of one of the most interesting efforts, due to von Karman, Heisenberg, and others, can be found in Birkhoff[1]; however, what was accomplished was not really a computation of S from basic theoretical principles, but rather the relating of S to other observable quantities connected with the "vortex street" that is formed behind the body.

Approaches to a Theoretical Analysis

It is by no means clear what should be the next step in the effort to calculate S, or at least the constants a and b, theoretically. It seems possible that a, which is just the limiting value of S as the viscosity approaches zero, could be calculated by finding periodic solutions of (**1.1**), (**1,2**), (**1.4**), and (**1.5**), and perhaps (**1.3**), since Euler's equations neglect viscosity; however, even this is not entirely clear. The difficulty is that if there is any viscosity at all, the condition (**1.2**) has to be replaced by the much stronger "no slip" boundary conditions:

$$V(y, t) = 0, \text{ for } y \text{ on } \Gamma \qquad (3.1)$$

But the argument that (**3.1**) should simply replace (**1.2**) is not entirely persuasive. It makes equally good sense to say that the behavior of a solution on the boundary should be described by a distribution describing the solution in an infinitely thin "boundary layer."

The preceding discussion shows that one has to be careful to distinguish between the case where viscosity is actually missing and the limiting properties of solutions as viscosity approaches zero. There is a similar effect with respect to compressibility, as recent work of Ebin[3] shows, so it may be misleading to limit consideration to incompressible flows.

There are still other reasons that it may be desirable to move away from planar, incompressible Euler flows. For instance, it may be too restrictive to view a 3-dimensional flow as planar even if the equations and boundary conditions are invariant under translation in the direction of some axis. Nothing assures a priori that a solution must have the same symmetry as the problem, although such assumptions are almost always made in practical problems. In fact, to the extent that solutions with a large amount of symmetry tend to be unstable, there is reason to give serious consideration to the possibility that asymmetric solutions should be sought.

My view is that it is not reasonable to expect to be able at this time to calculate S in general or to find more realistic exact solutions of the Euler equations (and still less the Navier-Stokes equations, which include viscous effects). The best hope would seem to lie in trying to develop methods of finding properties of solutions, or limits of solutions, without obtaining them explicitly. The remainder of the discussion here will be devoted to developing a method based on this idea.

The Geometric View

Suppose that $V(y, t)$ is a time-dependent vector field on E satisfying (**1.1**), (**1,2**), and (**1.4**), for $0 \leq t \leq T$, and define the one-parameter family of diffeomorphisms by the solution Y of the initial value problem

$$Y_t(x, t) = V(Y(x, t), t), \; Y(x, 0) = x \tag{4.1}$$

where the subscript denotes partial differentiation. (The solution will exist for $0 \leq t \leq T$ even though E is not compact because of (**1.1**).) Then for each fixed t, $Y(x, t)$ is a volume-preserving diffeomorphism of E and $Y(x, s + t) = Y(Y(x, s), t)$ whenever $0 \leq s, t, s + t \leq T$. Conversely, given such a family $Y(x, t)$, a time-dependent vector field V satisfying (**1.2**) and (**1.4**) is defined by setting

$$V(y, t) = Y_t(X(y, t), t) \tag{4.2}$$

where $X(y, t)$ is the inverse of the diffeomorphism $Y(x, t)$ for each fixed t. If the family Y arose from a vector field V by means of (**4.1**), (**4.2**) recovers V.

The Euler equation (**1.5**) can be derived by a variational argument.[7,11] If V and Y are related by (**4.1**), then V satisfies (**1.5**) if it satisfies the variational problem of providing a stationary value of the integral

$$\int_0^T \left\{ \int_E | Y_t(x, t) - V_\infty |^2 \, dx \right\} dt \tag{4.3}$$

with the "endpoints" $Y(x, 0)$ and $Y(x, T)$ fixed and subject to the constraints (**1.1**), (**1.2**), and (**1.4**).

The above calculus of variations argument can be cast in a somewhat more geometric form by regarding $Y(x, t)$ as a "curve" in the infinite dimensional manifold of volume-preserving diffeomorphisms of E (cf. references 4 and 5, where this approach is carried out for compact manifolds). From this point of view the calculus of variations problem becomes the geometric problem of finding paths of stationary length connecting the identity diffeomorphism to $Y(x, T)$.

This geometric point of view suggests that paths that are stationary, but not minimal, might be unstable, provided they are continued for a sufficiently long time. For instance the solutions of (**1.1**)–(**1.6**), discussed in the first section, do not necessarily minimize (**4.3**). The geometric view suggests further that the way to check for minimality is look for conjugate points along these solutions. This amounts to seeking solutions of the Jacobi equation[8] that vanish for $t = 0$ and $t = T$, where the "geodesic" is a solution of (**1.1**)–(**1.6**). It seems reasonable to conjecture that the smallest T for which a nontrivial solution of the Jacobi equation exists and vanishes for $t = 0$ and $t = T$ is related to the constant a of (**2.2**) by $2aT = 1$, that is, $2T$ is the limiting period of the vibrations induced by the flow as viscosity approaches zero.

My investigation from this point of view has not been completed, but the following sections are devoted to the first steps in carrying it out, that is, the derivation of the Jacobi equations for the context described and the discussion of the appropriate boundary conditions for them.

Notation and Formulas

The following notation (see appendix for more explicit formulas) will be needed: d = exterior differentiation; $*$ = Hodge operator; $\delta = *d*$ = metric transpose of d; $\Delta = d\delta + \delta d$ = Laplace operator, which for functions reduces to d; g = operator defined by the metric tensor that maps contravariant tensor fields to covariant tensor fields (lowering indices); g^{-1} = inverse of g (raising indices). Then: $\nabla f = g^{-1}\,df$ = gradient of a function f; $\nabla \cdot X = \delta g X$ = divergence of a vector field X; $\nabla \times X = g^{-1}\,dgX$ = curl of a vector field X (when the space is 2-dimensional X is a functon). Also, as above, $\nabla_X Y$ = covariant derivative of a vector field Y in the X direction (the Levi-Civita connection will always be used here) and $\nabla_X f$ = directional derivative of a function f in the X direction. Standard formulas, such as $d^2 = \delta^2 = 0$, will be used without comment.

The Jacobi Equation

Let E, X, Y, V be as above. If $W(y, t) = \nabla \times V(y, t)$, the Helmholtz-Kelvin Theorem[9,11] can be expressed as:

$$\frac{DW}{Dt} = W_t + \nabla_V W = 0 \tag{6.1}$$

(Here and below $D/Dt = \partial/\partial t + \nabla_V$ denotes the "material derivative.")

Now suppose the functions X, Y, V, etc. depend smoothly on a parameter ϵ in such a way that for each sufficiently small ϵ, a flow satisfying **(1.1)**, **(1.2)**, **(1.4)**, and **(1.5)** exists and, moreover, for $\epsilon = 0$, the flow also satisfies (**1.6**). Usually, the notation will not explicitly indicate the dependence on ϵ, since we shall only be concerned with the value $\epsilon = 0$ and the derivatives with respect to ϵ at this value. Denote these derivatives by Y' and V' and set $Z(y, t) = Y'(X(y, t), t)$. Now the equations that V' and Z satisfy will be derived.

Differentiating $Y_t(x, t) = V(Y(x, t), t)$ (cf. Equation 4.1) with respect to ϵ at $\epsilon = 0$ leads to $Y'_t(x, t) = V'(Y(x, t), t) + \nabla_{Y'} V(Y(x, t), t)$ or

$$Y'_t(X(y, t), t) = V'(y, t) + \nabla_Z \mathrm{V}(y, t) \tag{6.2}$$

Similarly, differentiating $Y'(x,t) = Z(Y(x, t), \mathrm{t})$ with respect to t leads to

$$Y'_t(X(y, t), t) = Z_t(y, t) + \nabla_V Z(y, t). \tag{6.3}$$

Equating the left sides of (**6.2**) and (**6.3**) gives

$$V' = Z_t + \nabla_V Z - \nabla_Z V = Z_t + [V, Z] \tag{6.4}$$

The equation of continuity (**1.4**) implies that

$$\nabla \cdot V' = 0 \tag{6.5}$$

Taking the divergence of both sides of (**6.4**) and using (**6.5**) leads after a short calculation to $D(\nabla \cdot Z)/Dt = (\nabla \cdot Z)_t + \nabla_V(\nabla \cdot Z) = 0$, in particular,

$$\nabla \cdot Z = 0, \text{ for all } t \text{ if } \nabla \cdot Z = 0 \text{ for } t = 0. \tag{6.6}$$

The assumption that $\nabla \cdot Z = *d*gZ = 0$ means that the 1-form $*gZ$ is closed. It is natural to require that the perturbations leave the flow tangent to the boundary, hence V' and Z are required to be tangent to Γ, from which it follows that $*gZ$ must be exact, say, $*gZ = dF$, or

$$gZ = *dF \tag{6.7}$$

Setting $W' = \nabla \times V'$ and differentiating (**6.1**) with respect to ϵ leads to

$$\frac{DW'}{Dt} = W'_t + \nabla_V W' = 0 \tag{6.8}$$

since, when $\epsilon = 0$, $W = \nabla_{V'} W = 0$.

The equations (**6.4**), (**6.5**), (**6.7**), and (**6.8**), together with $W' = \nabla \times V'$ are, in effect, the Jacobi equation for variations along the flow determined by V; however, it is more convenient to eliminate some of the variables to obtain an equation in F alone. To do this, note that $g[V, Z] = *d\nabla_V F$ follows from $\nabla \cdot V = 0$ and (**6.7**) by a straightforward computation in local coordinates, hence (**6.4**) implies that

$$gV' = \frac{*dDF}{Dt} = *d(F_t + \nabla_V F) \tag{6.9}$$

Applying $*d$ to both sides of (**6.9**) and using (**6.8**) leads to the desired relation

$$\frac{D}{Dt} \circ \Delta \circ \frac{DF}{Dt} = 0. \tag{6.10}$$

There are also boundary conditions that F must satisfy in addition to (**6.10**). Since Z must be tangent to Γ, it is necessary that

$$F = \text{constant on } \Gamma. \tag{6.11}$$

Moreover (**1.1**) leads to

$$dF(y, t) \to 0, \text{ as } |y| \to \infty \tag{6.12}$$

If one takes the energy constraint (**1.3**) seriously, it is also to be expected that

$$\int_D |\nabla V(y, t)|^2 \, dy \leqq \text{const.} < \infty. \tag{6.13}$$

Thus (**6.10**) together with the boundary conditions (**6.11**), (**6.12**), and possibly (**6.13**), is the analog of the Jacobi equation. A time $T > 0$ corresponds to a conjugate point along the unperturbed flow provided there is a nontrivial solution satisfying

$$F(x, 0) = F(x, T) = \text{const.} \tag{6.14}$$

Existence of Solutions

It is easy to see that solutions of (**6.10**)–(**6.14**) exist. In fact, let Ω_0 be a real valued continuous function defined on D and tending to zero at infinity in a sufficiently strong sense (e.g., let Ω_0 have compact support). Then define $\Omega(y, t)$ as the solution of $D\Omega/Dt = 0$, $\Omega(x, 0) = \Omega_0(x)$, that is, $\Omega(y, t) = \Omega_0(X(y, t))$. Let $\Lambda(y, t)$ be the solution of Poisson's equation $\Delta\Lambda = \Omega$, $\Lambda(y, t) \to 0$ as $|y| \to \infty$, and $\Lambda(y, t) = 0$ on Γ, which will exist if Ω_0 tends to zero strongly enough at infinity. The desired solution of (**6.10**)–(**6.14**) is obtained by letting F be a solution of $DF/Dt = \Lambda$, that is, if we set $G(x, t) = \int_0^t \Lambda(Y(x, s), s)\,ds$, then $F(y, t) = G(X(y, t), y)$. We omit the details of the proof in which the strong tending of Ω_0 to zero implies that (**6.13**) and (**6.14**) are satisfied.

It is not presently known if there are any conjugate points along the geodesics defined by the solutions of (**1.1**)–(**1.6**), or what is the same thing, if there exist any nontrivial solutions of (**6.10**)–(**6.15**). Even if this question were resolved it would still remain to compute the smallest possible value of T and see if it bears the expected relationship,

$$\lim_{R \to \infty} S = a = \tfrac{1}{2}T$$

to the Strouhal number, as described in The Geometric View.

References

1. Birkhoff, G. & E. H. Zarantonello. 1957. Jets, Wakes, and Cavities. Academic Press, New York, N.Y.
2. Birkhoff, G. 1956. Hydrodynamics: A Study in Logic Similitude and Fact. Princeton University Press, Princeton, N.J.
3. Ebin, D. G. 1975. Motion of a slightly compressible fluid. Proc. Nat. Acad. Sci. U.S.A. **72:** 539–542.
4. Ebin, D. G. 1971/72. Espace de métriques Riemanniennes et mouvement des fluides via variétés d'applications. Lect. Notes Ecole Polytech. Université Paris VII.
5. Ebin, D. G. & J. Marsden. 1970. Groups of diffeomorphisms and the motion of an incompressible fluid. Ann Math. **92:** 102–163.
6. Goldstein, S. 1965. Modern Developments in Fluid Dynamics. Dover, New York, N.Y.
7. Lichtenstein, L. 1929. Grundlagen der Hydromechanik. Springer-Verlag, Berlin.
8. Milnor, J. 1963. Morse Theory. Princeton University Press, Princeton, N.J.
9. Marsden, J. 1974. Applications of Global Analysis in Mathematical Physics. Publish or Perish, Boston, Mass.
10. Rayleigh, Lord. 1915. Aeolian Tones. Philosophical Magazine. **29:** 433–444.
11. Sommerfeld, A. 1950. Mechanics of Deformable Bodies. Academic Press, New York, N.Y.

APPENDIX

The invariantly defined operators, discussed under NOTATION AND FORMULAS, take on particularly simple forms if one works with rectangular coordinates in a Euclidean space. In that case g and g^{-1} are represented by identity matrices of the appropriate size and the Christoffel symbols vanish. Some of the most important operators take the following form for functions $f(x, y)$, vector fields $X = X_1\partial/\partial x + X_2\partial/\partial y$ and Y, and 1-forms $\omega = adx + bdy$, defined on open subsets of IR^2.

$$df = (\partial f/\partial x)dx + (\partial f/\partial y)dy$$
$$\Delta f = \partial^2 f/\partial x^2 + \partial^2 f/\partial y^2$$
$$\nabla_X Y = (X_1\partial Y_1/\partial x + X_2\partial Y_1/\partial y)\partial/\partial x + (X_1\partial Y_2/\partial x + X_2\partial Y_2/\partial y)\partial/\partial y$$
$$gX = X_1 dx + X_2 dy$$
$$g^{-1}\omega = a\partial/\partial x + b\partial/\partial y$$
$$*\omega = ady - bdx$$
$$*f = fdx \wedge dy$$
$$\nabla \times X = \partial X_2/\partial x - \partial X_1/\partial y$$
$$\nabla \cdot X = \partial X_1/\partial x + \partial X_2/\partial y.$$

It is important to remember that these formulas have to be modified even if polar coordinates are employed in IR^2.

TREES AND BALL GAMES

Raymond M. Smullyan

Department of Mathematics
Herbert H. Lehman College
City University of New York
Bronx, New York 10468

THE BALL-GAME THEOREM

We consider the following one-man game: We have an infinite supply of (pool) balls numbered 1, an infinite supply of balls numbered 2, and for each positive integer *n*, we have infinitely many balls numbered *n*. We shall sometimes refer to the number on a ball as the *rank* of the ball. We shall imagine all these infinitely many balls as lying on an (infinite) floor. On a table is lying a box with infinite capacity. In this box is lying a finite number of pool balls (each ball in the box has a number, just like the balls on the floor). Now, the rule of the game is this: At any stage, the player may remove from the box any ball and then replace it with any finite number of balls of lower rank. For example, he may throw out a ball of rank 57 and replace it with a billion balls of rank 56, or rank 48, or some of one rank and some of another, providing all the ranks are less than 57. If the player throws out a ball of rank 1, he can't replace it with anything. If ever the box becomes empty, the player loses.

The problem is whether the player must eventually lose, or whether (with sufficient ingenuity) he can keep the process going forever. On the one hand it might seem that if there is at least one ball of rank higher than 1, we can get as large a finite number of balls in the box as we please, hence we should be able to keep the process going forever. On the other hand, it might seem that even a single ball of a given rank *n* is somehow "worth more" than *any* finite number of balls of lower rank; hence every move we make is somehow "worsening" our situation. Which is really the case?

THEOREM 1 (Ball-Game Theorem). The process must eventually terminate.

One can prove this in several ways; one way is by the following induction argument.

If all balls in the box are of rank 1, then we obviously have a losing game. Suppose the highest rank of any ball in the box is 2. Then we have at the outset a finite number of 2's and a finite number of 1's. We can't keep throwing away 1's forever, hence we must sooner or later throw out one of our 2's. Then we have one less 2 in the box (but possibly many more 1's than we started with). Again, we can't keep throwing out 1's forever, so we must sooner or later throw out another 2. We see that after a finite number of steps, we must throw away our last 2, and then we are back to the situation in which we have only 1's. We already know this to be a losing situation. This proves that the process must terminate if the highest rank present is 2. Now, what if the highest rank is 3? We can't keep throwing away just balls of rank ≤ 2 forever (we just proved that!); hence we must sooner or later throw out a 3. Then again we must sooner or later throw out another 3, and so we must eventually throw out our last 3. This then

0077–8923/79/0321–0086 $01.75/1 © 1979, NYAS

reduces the problem to the preceding case when the highest rank present is 2, which we have already solved.

This obviously suggests the pattern of the induction argument, whose formal details we leave to the reader. (If formalized, the argument will be seen to involve not only an induction on the number *n*, which is the highest rank present, but also on the number *k* of balls present of rank *n*.)[1]

Another solution of the problem is via König's Lemma, which we give in the next section. In the concluding section of this paper, we show how this ball problem can provide an alternative proof of König's Lemma.

KÖNIG'S LEMMA

We shall define a *tree* T as a collection of the following items:

(1) A set S of elements called *points*.
(2) An assignment to each point x a positive integer $l(x)$ called the *level* of x.
(3) A relation xRy between elements of S, which we read "y is a successor of x," or "x is a predecessor of y." This relation must obey the following conditions:

c_1. There is a unique point b_1 of level 1. This point we call the *origin* of the tree.

c_2. Every point of the tree other than the origin has a unique predecessor; the origin has no predecessors.

c_3. For any points x,y if y is a successor of x, then $l(y) = l(x) + 1$.

We call a point x an *endpoint* of T if it has no successors. By a *path* we mean any finite or denumerable sequence of points, beginning with the origin, which is such that each term of the sequence, except the origin, is the successor of the preceding term. By a *maximal path* or branch of T, we mean a path whose last term is an endpoint or a path which is infinite.

We call a set A of points *iterative* if it satisfies the following two conditions:

I_1. A contains at least one point.

I_2. Every point x in A has at least one successor in A.

THEOREM 2. The following two conditions are equivalent:

c_1. There exists at least one iterative set.

c_2. T contains at least one infinite branch.

Proof: It is trivial that c_2 implies c_1, for if T contains at least one infinite branch, then the set of points on this branch is obviously iterative.

Now, suppose there is at least one iterative set A. This A contains at least one point x. Let $(a_1, \ldots, a_k, x)$ be a path ending in x (k may be 0). By I_2, x has at least one successor x_1 in A, which in turn has at least one successor x_2 in A, and so forth. Thus we have the infinite branch $(a_1, \ldots, a_k, x, x_1, x_2, x_3, \ldots, x_n, \ldots)$.[2] □

We call a point y a *descendant* of a point x if either $y = x$ or there is at least one branch in which y occurs as a later term than x. (Thus, the descendants of a point x

consist of x, the successors of x, the successors of the successors of x, and so forth.)

We call x a *good* point of T if it has infinitely many descendants and a *bad* point if it has only finitely many descendants.

LEMMA. If x is good and if x has only finitely many successors, then at least one successor of x is good.

Proof: Suppose x has only finitely many successors $x_1, \ldots, x_k$. Suppose each of these successors were bad. Let N_1 be the number of descendants of $x_1, \ldots$ and N_k the number of descendants of x_k. Thus the $N_1, \ldots, N_k$ are all finite. Then $N_1 + \ldots + N_k + 1$ is the number of descendants of x, so x must be bad. Thus if x is good and x has only finitely many successors $x_1, \ldots, x_k$, then at least one of the $x_1, \ldots, x_k$ must be good. □

Now, let us call a tree K a *König* tree if each point has only finitely many successors.

THEOREM 3 (König's Lemma). In a König tree K, if there are infinitely many points on K, then K contains at least one infinite branch. Equivalently, in a König tree K, if all the branches are finite, then the whole tree is finite (i.e., it contains only finitely many points.)[3]

Proof (After König): Let K be an infinite König tree. Since there are infinitely many points on the tree and all points are descendants of the origin, then the origin is good, hence there is at least one good point. Also, by the above lemma, every good point has at least one good successor. These two facts mean that the set of all good points is iterative. Result then follows from THEOREM 2. □

Application to the Ball-Game Problem

Now let us return to the ball-game problem of the first section. To each ball game we associate a tree as follows: We take for the "origin" of the tree any element whatsoever (its nature will have no significance). Its "successors" shall be the balls which are in the box at the outset of the game. When we "replace" any ball by other balls (of lower rank) we take these other balls as the "successors" of the ball we replace. This is the tree associated with the process of the game.

Clearly the tree is a König tree, since each ball is replaced by only finitely many balls (and also there are only finitely many balls initially in the box). Also each branch of the tree must be finite (because any successor of a ball has lower number than the ball!) Then by König's Lemma, the tree has only finitely many points. However, the points of the tree are precisely those balls which ever enter the box (either initially or as replacements). Therefore only finitely many balls are ever in the box. If the game goes on infinitely, then infinitely many balls would have to enter the box. Therefore the game must be finite.

We have just seen how the Ball-Game Theorem (THEOREM 1) can be obtained as a special case of König's Lemma. Actually, König's Lemma for denumerable trees is really no more general than the Ball-Game Theorem, as we shall now demonstrate by showing that every König tree whose branches are all finite is in fact the tree of some ball game.

For any set B of points of T, we shall call B *inductive* if, for every point x, B contains all successors of x and x, as well. (An inductive set automatically contains all endpoints x of a tree, since it vacuously contains all successors of x and x has no successors.

THEOREM 4 (A generalized induction principle).[4] For any tree T (not necessarily a König tree), if all branches are finite, then the only inductive set is the set of all points of the tree.

Proof: Let B be an inductive set. Let C be the set of all points of T which are not in B. Every element x of C must contain at least one successor which is in C because if no successors of x were in C, they would all be in B, which would mean that x was in B (since B is inductive), contrary to the assumption that x is in C. Therefore C has the property of I_2 of our definition of "iterative" (see paragraph immediately preceding THEOREM 2). If C were nonempty, then C would be iterative, which, by THEOREM 2, would imply that T has at least one infinite branch. Therefore if T has no infinite branches, then C must be empty, which means that B is the set of all points of the tree. □

THEOREM 4 means that for any tree T whose branches are all finite, given any property P which we wish to show holds for all points of the tree, it suffices to show that for every point x, if P holds for all successors of x, then P also holds for x (because this means that the set of all points for which the property does hold is an inductive set).

THEOREM 5. Given any König tree whose branches are all finite, it is possible to assign to each point x a natural number $r(x)$ (which we will call the *rank* of x) such that the following two conditions fulfilled for every point x.

R_1. If x is an endpoint, then $r(x) = 1$.

R_2. If x is not an endpoint, then x has higher rank than any of its successors.

Proof: Let K be a König tree in which all branches are finite. For each point a, let D_a be the set of descendants of a. Let us say that a is *passable* if it is possible to assign natural numbers to all elements of D_a in such a manner that conditions R_1 and R_2 hold for every x in D_a—also, let us call any such assignment of numbers to all elements of D_a a *passable assignment*. We are to prove that the origin of the tree is passable. To do this it of course suffices to show that every point of the tree is passable. And to do this, it suffices (by THEOREM 4) to show that, for every point a, if all successors of a are passable, then a is passable. So let us assume that all successors of a are passable.

If a is an endpoint, then a is certainly passable, since a is then the only member of D_a; so we assign a the number 1.

Suppose a is not an endpoint. Now a has only finitely many successors $a_1, \ldots, a_k$. We are assuming that, for each $i \leq k$, the point a_i is passable. For each $i \leq k$, let A_i be a passable assignment for the set D_{a_i}. We then assign numbers to all elements of D_a as follows: For each x in D_a, other than a itself, x belongs to exactly one of the sets $D_{a_1}, \ldots, D_{a_k}$. We take the set D_{a_i} to which x belongs, and assign to x the same number assigned to x by A_i. We have now assigned numbers to all elements of D_a, except for a. Let $n_1, \ldots, n_k$ be the numbers already assigned to $a_1, \ldots, a_k$. Then assign to a any number greater than each of the $n_1, \ldots, n_k$. We have now assigned numbers to all elements of D_a, and this assignment is clearly passable. Thus a is a passable point. This concludes the proof. □

It follows from THEOREM 5 that every König tree whose branches are all finite can be made from the tree of some ball game.[5] If the tree has infinitely many points, then the ball game can never terminate (because infinitely many balls will enter the box), which is contrary to THEOREM 1. Therefore the tree must be finite. This then provides an alternative proof that a König tree which is infinite must contain at least one infinite branch.

NOTES

1. As pointed out by the referee, the following alternative proof, using ordinal numbers is a simple one:

   If the number of balls of rank j is denoted b_j, associate with each state of the box the ordinal

   $$b_n\omega^n + b_{n-1}\omega^{n-1} + \ldots + b_1\omega$$

   where n is the greatest rank of the balls in the box. Then each move in the game produces a smaller corresponding ordinal. Hence, since there cannot be an infinite decreasing sequence of ordinals, the game must end in a finite number of moves.
2. Our proof, of course, used the Axiom of Choice. However, the only trees we shall need to consider for our applications are those in which all the points come from some denumerable set—in this case the Axiom of Choice is not needed.
3. When I said "equivalently" I meant *equivalently* in classical logic. In intuitionistic logic, the two formulations are not equivalent.
4. This theorem is well known; I include its proof in order to make this paper self-contained.
5. Once the points of the tree are numbered as in THEOREM 5, we can think of the points as numbered balls, and the tree is then the tree of any ball game such that: (1) the origin is the one and only ball initially in the box, and (2) each ball gets replaced by its successors on the tree.

THE FOUR COLOR PROBLEM

Joel G. Stemple

Department of Mathematics
Queens College
City University of New York
Flushing, New York 11367

For almost a century, the Four Color Problem has been one of the best known unsolved problems in mathematics. Its status has changed recently with the proof announced by Appel and Haken (of the University of Illinois). In this paper, the development of the problem will be traced from its beginning, through various attempts at solving it, culminating in a brief explanation of the techniques that were employed in the solution.

If one is given a *map* consisting of a number of countries, it is normal procedure to color the map so that whenever two countries have a common border, they receive different colors (if two countries meet at just a point, they are permitted to have the same color). The question then arises: what is the smallest number of colors needed to color the countries of any map in the indicated manner? The principal attempts to solve this problem have come from the area of mathematics known as the theory of graphs. In fact, the growth of this field is due mainly to the problem under consideration.

DEFINITION. A *graph G* consists of a set of designated points, called *vertices,* and a set of curved or straight lines, called *edges,* connecting pairs of vertices. A graph is *planar* if it can be drawn in the Euclidean plane so that two edges may only touch at a common vertex.

The map referred to above can now be thought of more formally as a planar graph in which the vertices are the points at which three or more countries come together, and the edges are the common portions of the boundary (see FIGURE 1). The countries of the map become what are called the *faces* of the graph. When we consider the faces, we also include the external face of the graph. Thus, in FIGURE 1, there are five faces.

We assume that our graph G satisfies certain minor restrictions when we attempt to color it; these include:

(i) No face has a boundary edge with itself (such an edge is called a *separating edge*).
(ii) No face is surrounded by just a single edge.
(iii) The graph is connected (i.e., it does not consist of two or more totally separate pieces).

In FIGURE 2, edge E violates condition (i), E' violates (ii), and the entire graph consists of two parts, thus violating (iii).

All attempts to color such graphs have resulted in the need for four or fewer colors. This leads to the following:

0077-8923/79/0321-0091 $01.75/1 © 1979, NYAS

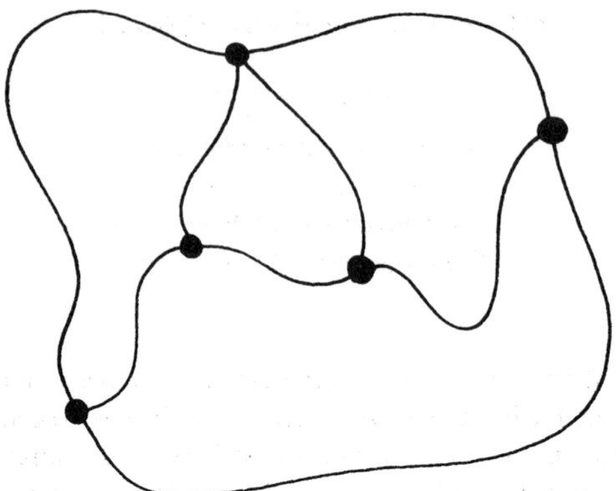

FIGURE 1.

FOUR COLOR CONJECTURE. The faces of a planar graph can always be colored with no more than four colors in such a way that faces with a common edge receive different colors.

The *Four Color Problem* refers to the problem of determining whether this conjecture is true or false. In attempting to solve this problem, mathematicians have discovered further restrictions that may be placed on the graph G. We shall proceed to investigate some of these.

DEFINITION. If v is a vertex of G, then the *degree* (or *valence*) of v, denoted $\rho(v)$, is the number of edges of G having v as an endpoint. In particular if $\rho(v) = 3$ for all vertices, then G is called *trivalent*.

Let G be a planar graph satisfying the restrictions indicated above. We may then add the following assumptions:

(iv) $\rho(v) \neq 1$ (if $\rho(v) = 1$, we violate restruction (i)).

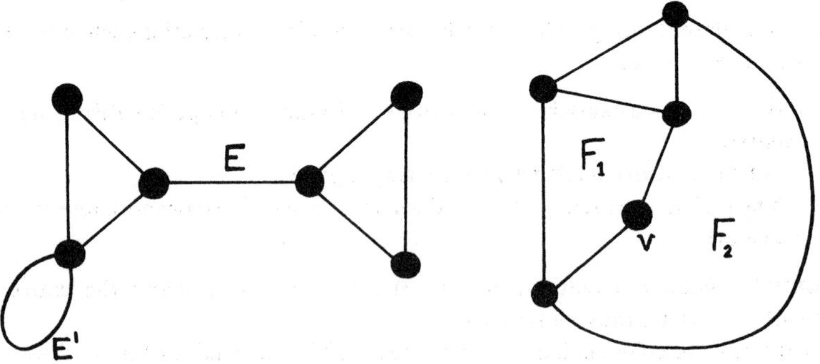

FIGURE 2.

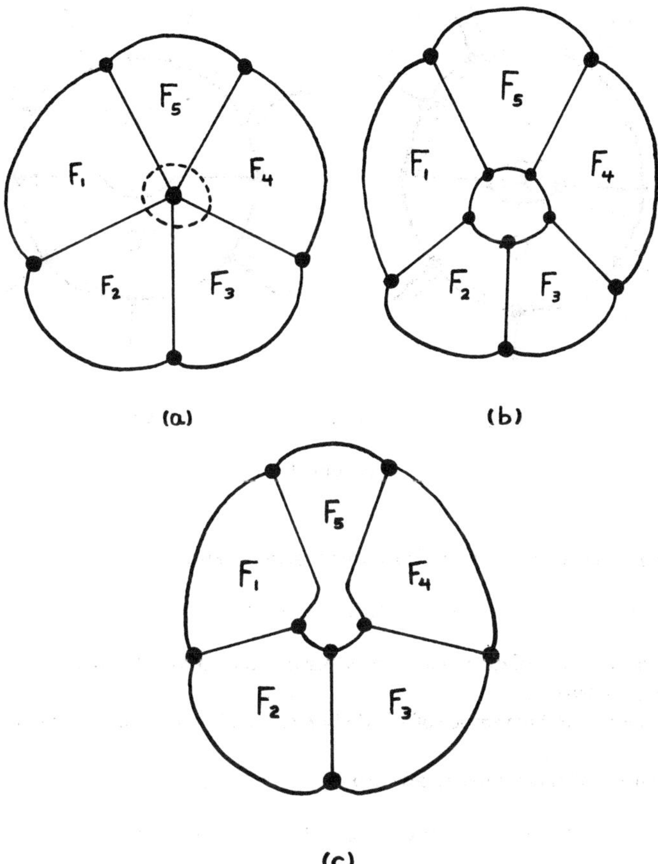

FIGURE 3.

(v) $\rho(v) \neq 2$ ($\rho(v) = 2$ is redundant. In FIGURE 2, faces F_1 and F_2 are separated by two edges as a result of v. We obtain the same information by replacing the two edges between F_1 and F_2 by a single edge.)

(vi) $\rho(v) = 3$ for all vertices (We illustrate how to eliminate a vertex of degree 5. In FIGURE 3a draw a small circle around v (the vertex at the center); then remove v and all portions of the edges inside the circle. We have thus replaced a vertex with degree greater than 3 by several vertices of degree 3 (see FIGURE 3b). This procedure has resulted in one extra face in the graph. However, a slight modification of this method results in a trivalent graph with the same number of faces as the original graph (See FIGURE 3c). Note that any method of coloring the graph in FIGURE 3c will also work in FIGURE 3a).

DEFINITION. The number of vertices, edges, and faces in G are denoted by υ_v, υ_e, and υ_f, respectively.

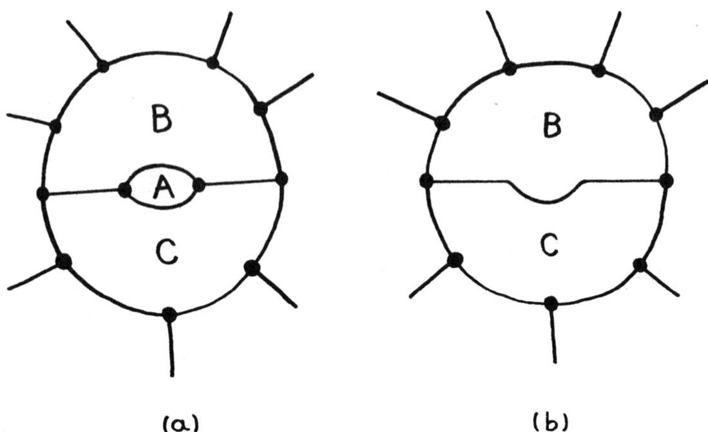

FIGURE 4.

Euler's formula for a connected planar graph states that

$$v_v + v_f - v_e = 2$$

For the duration of our discussion, we assume that G is a planar graph satisfying restrictions (i)—(vi).

We let ϕ_k denote the number of k-sided faces in G. By restriction (ii), we know that $\phi_1 = 0$.

The number of faces in G is given by:

$$v_f = \phi_2 + \phi_3 + \phi_4 + \phi_5 + \phi_6 + \phi_7 + \phi_8 + \ldots . \quad \textbf{(1)}$$

Each vertex is on three faces. Hence, if we add together the total number of vertices on each face, we obtain three times the number of vertices:

$$3v_v = 2\phi_2 + 3\phi_3 + 4\phi_4 + 5\phi_5 + 6\phi_6 + 7\phi_7 + 8\phi_8 + \ldots . \quad \textbf{(2)}$$

Similarly, since each edge is on two faces, we have:

$$2v_e = 2\phi_2 + 3\phi_3 + 4\phi_4 + 5\phi_5 + 6\phi_6 + 7\phi_7 + 8\phi_8 + \ldots . \quad \textbf{(3)}$$

Writing Euler's formula as

$$12 = 6v_v + 6v_f - 6v_e \quad \textbf{(4)}$$

and substituting (**1**), (**2**), and (**3**) into (**4**) yields:

$$12 = 4\phi_2 + 3\phi_3 + 2\phi_4 + \phi_5 - \phi_7 - 2\phi_8 - \ldots . \quad \textbf{(5)}$$

Since the left-hand side of (**5**) is positive, there must be *positive* terms on the right. This leads to the next result.

THEOREM 1. If G is a trivalent, connected, planar graph with no separating edge, then some face of G has five or fewer sides.

To try to prove the Four Color Conjecture, we assume it to be false (and hope to show this is impossible). Then there must be a *minimal* graph G requiring five colors—"minimal" is used in the sense that any graph containing fewer faces than G will require no more than four colors. We can now show that in a minimal graph certain types of faces cannot exist.

THEOREM 2. If G is a minimal counterexample to the Four Color Conjecture then G contains no 2-, 3-, or 4-sided faces.

Proof: If G contains a 2-sided face, then the portion of G containing this face appears in FIGURE 4a. We remove one of the edges on the two sided face and, after also eliminating the vertices of degree 2 as indicated in restriction (v), we have a new graph G' with fewer faces (see FIGURE 4b). By our definition of minimal graph, the smaller graph G' can be colored in four colors. We now return to G and give every face the

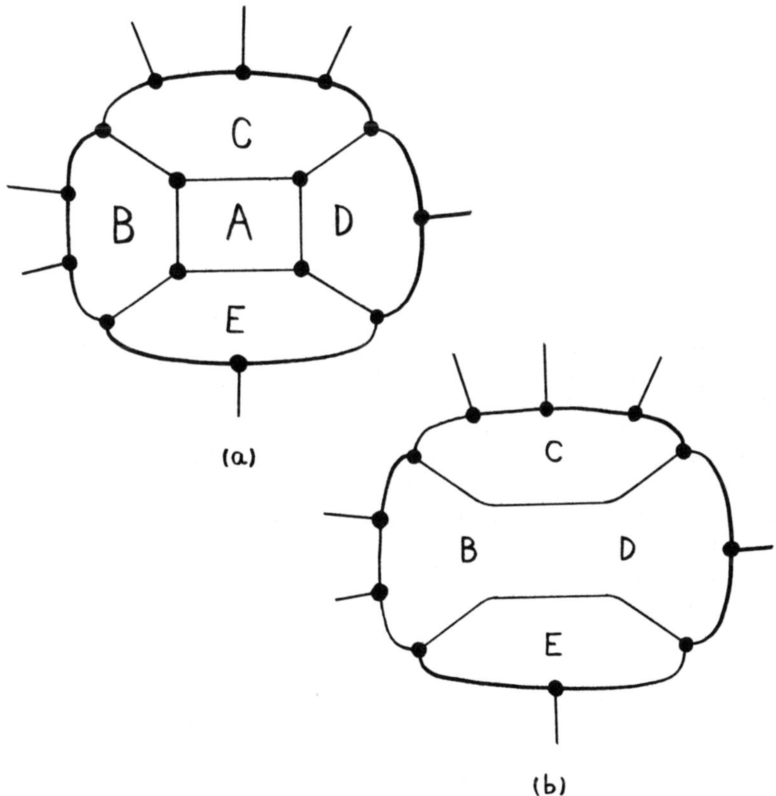

FIGURE 5.

same color that it had in G'. This leaves only face A, the two sided face, with no color. But since A only touches faces B and C we can assign to A one of the two colors not used by B and C. Hence G is now completely colored with four colors, contradicting the assumption that G is a minimal counterexample to the Four Color Conjecture.

In similar manner, we can show that there can be no 3-sided faces in G.

To eliminate 4-sided faces, we only vary our method slightly. Referring to FIGURE 5a, we observe that at least one pair of opposite faces—either B and D or instead C and E—do not touch elsewhere in the graph. We shall assume that B and D do not touch. Then eliminate the edges between A and B and between A and D, obtaining the graph G' shown in FIGURE 5b. Since this graph has fewer faces, it can be four colored. We return to G, using the same colors. Then B and D are both assigned the same color in G (they had the same color in G' since they were combined into a single face in G'). Hence, even if C and E have different colors, there is still a color left over to assign to A and thus G is colored in four colors.

This procedure, unfortunately, does not work if we try to eliminate 5-sided faces. However, if five colors had been allowed from the beginning (and our proof by contradiction started with the assumption that the minimal graph G required six colors) then we could show (employing a method similar to that used for a 4-sided face) that 5-sided faces can be eliminated. However, eliminating all 2-, 3-, 4-, and 5-sided faces contradicts THEOREM 1. Thus we have the following:

THEOREM 3. The faces of a planar graph can always be colored with five or fewer colors.

In 1852 Professor Augustus de Morgan of University College in London wrote a letter to Sir William Rowan Hamilton of Trinity College. Included in this letter is the first known mention of the Four Color Problem. Professor de Morgan states that a student of his (Francis Guthrie) asked him this question and that he did not see an immediate answer. The problem caused no stir at that time. It was not until Cayley posed the same question at a meeting of the London Mathematical Society in 1878 that mathematicians became interested. A year later, Kempe published a "proof" and the matter was thought to be settled. In 1890, Heawood wrote an article showing that Kempe's proof was incorrect and at the same time advancing some new thoughts about the problem. Mathematicians now felt challenged, and the Four Color Problem gained new importance.

Kempe's attempt to prove the Four Color Conjecture started with the method we used to justify THEOREM 3. The principal difference was his treatment of the 5-sided faces. Here, he introduced an ingenious device involving chains of regions (these have since been termed "Kempe chains"). Although his proof was false, the chain technique has survived and has been correctly used to obtain further restrictions on the graph G. For a very readable description of Kempe's proof and an explanation of the fallacy, see Reference 3.

The basic idea in most attempts at proving the Four Color Conjecture has been to work with configurations that contain more than just one 5-sided face. A configuration is said to be *reducible* if it cannot exist in a minimal graph requiring five colors. Thus, we have already shown that a 4-sided face is reducible (so are 2- and 3-sided faces). By using Kempe chains and considering a number of cases, it can be shown that a 5-sided face which touches three consecutive 5-sided faces is reducible (see FIGURE 6a). Another example of a reducible configuration is given in FIGURE 6b.

Since 2-, 3-, and 4-sided faces are not possibilities in a minimal graph G requiring five colors, formula (5) becomes

$$12 = \phi_5 - \phi_7 - 2\phi_8 - 3\phi_9 - \dots. \tag{6}$$

It follows that $\phi_5 \geq 12$ and hence the number of faces is at least 12. Using the fact that the configuration in FIGURE 6a is reducible, it follows that not all faces can be 5-sided. Thus $v_f = \phi_5 + \phi_6 + \phi_7 + \dots > 12$.

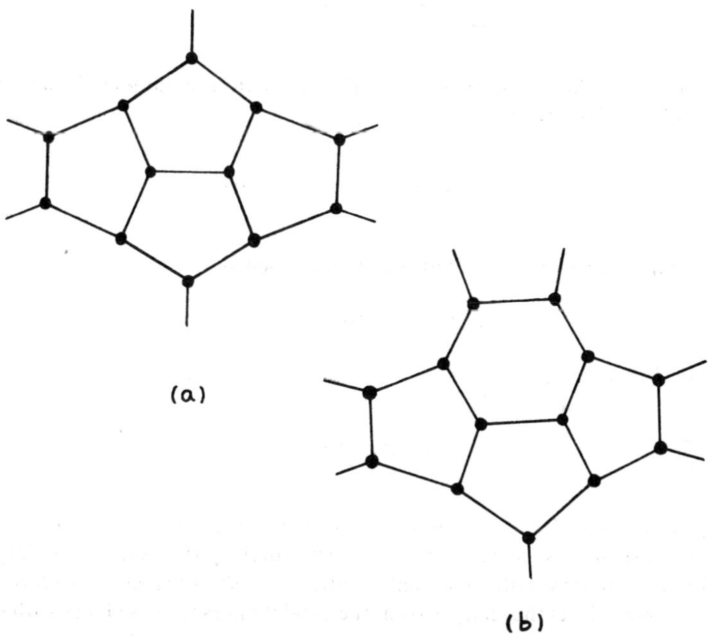

FIGURE 6.

As more reducible configurations were discovered, this number was increased. In 1922, Franklin showed that a minimal graph requiring five colors must satisfy $v_f \geq 26$. This was improved to $v_f \geq 27$ (Reynolds, 1926), $v_f \geq 32$ (Franklin, 1938), and $v_f \geq 36$ (Winn, 1940). Winn's result remained for twenty-eight years until Oystein Ore and I, using three new reductions, showed that $v_f \geq 40$.[5] Professor Ore stated at that time that he felt this improved result would result in renewed and more determined attempts to solve the Four Color Problem. In 1974, Walter Stromquist[7] showed that $v_f \geq 52$ and finally, in 1976, Kenneth Appel and Wolfgang Haken announced that they had proven the Four Color Conjecture.

Appel and Haken's announcement was greeted with skepticism. This is not surprising since, over the years, many "proofs" of this conjecture have been given and all were eventually found to be incorrect. To complicate matters, many of the new reducible configurations used by Appel and Haken were found by computer. All told, in proving the result they used approximately 1200 hours of computer time and found close to 2000 reducible configurations. By now, their paper[1] has been carefully checked by other mathematicians and their proof is generally accepted (although there remain some mathematicians who still question either the proof's validity or the propriety of using a computer to help prove a result).

The idea of the Appel and Haken proof follows. Note that the coefficients of the ϕ_k terms in (**6**) are of the form $(6\text{-}k)$. Multiplying (**6**) by 60 then yields

$$\sum_{k=5}^{\infty} 60\,(6 - k)\,\phi_k = 720 \tag{7}$$

If F is a face in the graph, then we let $\rho^*(\mathrm{F})$ denote the number of edges on that face. Then (**7**) can be written

$$\sum_{F \subset G} 60\,(6 - \rho^*(F)) = 720 \tag{8}$$

The *initial charge* on F, denoted $q_0(F)$, is defined by

$$q_0(F) = 60\,(6 - \rho^*(F)) \tag{9}$$

Thus, (**8**) becomes

$$\sum_{F \subset G} q_0(F) = 720 \tag{10}$$

Observe that $q_0(F) > 0$ if and only if $\rho^*(F) = 5$. For $\rho^*(F) \geq 7$, $q_0(F) < 0$.

Since G is assumed to be a minimal counterexample to the Four Color Conjecture, G may not contain any reducible configurations. A *discharging procedure* is then established, which shifts the charges from the 5-sided faces to those faces with seven or more sides. Letting $q(F)$ denote the new charge assigned to each face, the discharging is done in such a manner that the total charge remains unchanged; that is,

$$\sum_{F \subset G} q(F) = \sum_{F \subset G} q_0(F) = 720 \tag{11}$$

Appel and Haken show how, in a graph with no reducible configurations, the charges can be shifted so that all faces end up with charges that are zero or negative. But then the sum of such charges cannot add up to a positive number; i.e., 720 in (**11**). It follows that no such graph can exist, and hence there is no minimal counterexample to the Four Color Conjecture. This establishes the main theorem:

THEOREM 4 (*The Four Color Theorem*). The faces of a planar graph can always be colored with four or fewer colors.

In working with the computer, it was an impossible task (due to the amount of time involved) to try to find all reducible configurations. Also, a configuration might be reducible even if the computer program does not succeed. The Appel and Haken proof took several years. They started with a discharging method and then, if something ended up with a positive charge, tried to show that it was part of a reducible configuration. If they could not do this, then the discharging method had to be modified so that the positive charge was removed. They went back and forth in this manner, looking at approximately 10,000 different cases in which a face could have positive charge.

The final discharging procedure involved hundreds of different possibilities. Two illustrations of how it works are shown in FIGURES 7 and 8. The borders of the 5-sided

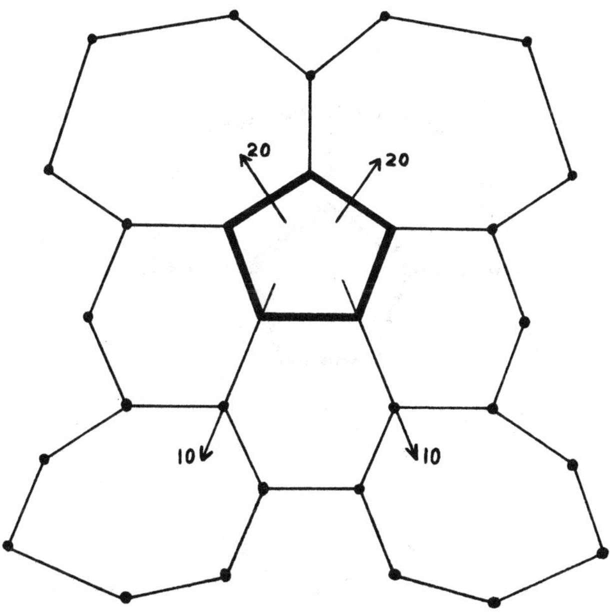

FIGURE 7.

faces (the ones with positive charge) are darkened and the arrows with numbers next to them show how much charge is transferred and to which face. Thus in FIGURE 7, the 5-sided face, which originally had a charge of 60, is left with a charge of zero. The four 7-sided faces started with charges of −60. Two end up with charges of −40 and two with −50. (The situation in FIGURE 7 is actually over simplified. Whether this specific method of discharging would have been used depends also upon whether other 5-sided faces touch the faces shown.) In FIGURE 8, only one of the 5-sided faces is discharged. To discharge the other two requires looking at the remaining faces they touch and this involves many separate cases.

The complexity of the arguments involved can be seen if we imagine what would happen if the configurations in FIGURES 7 and 8 were combined by assuming that the 7-sided face at the bottom of FIGURE 8 is the same as the 7-sided face on the top right in FIGURE 7. Then a total charge of 20 + 60 would have been transferred to this face, leaving it with a charge of +20. Thus the face is "overcharged." What is then necessary is to check whether the combined configuration is reducible or else to modify the discharging procedure that is used here (this is what was actually done).

Appel and Haken[1] explain why it seems reasonable that any proof of the Four Color Theorem that involves reducible configurations and some sort of discharging procedure will be long and complicated. They admit that their proof can be shortened somewhat (in fact, their initial announcement in 1976 said that 1936 reducible configurations had been used,[6] but, by October, 1977 they had reduced this number to 1482[2]), but then state that it will never be able to be made into a neat and elegant mathematical proof.

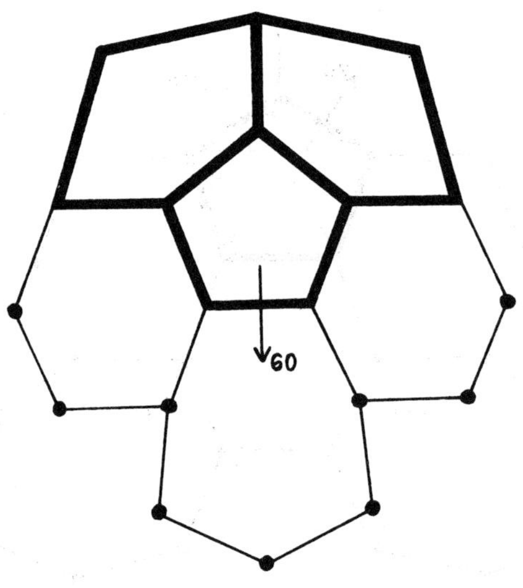

FIGURE 8.

In our discussion, we worked with the original face coloring formulation of the Four Color Conjecture. An equivalent method involves the "dual graph" which is defined as follows:

DEFINITION. Let G be a planar graph. Form a new graph G' by placing inside each face of G a new vertex and then drawing an edge between two vertices of G' if and only if they lie inside faces of G that have a common border. G' is called the *dual graph* of G. (We can think of G' as being formed by letting its vertices be the capitol cities of the countries to be colored and letting its edges be roads that join the capitols of neighboring countries.)

The effect of forming G' from G is to reverse the roles of the faces and vertices. If all vertices of G have degree 3, then all faces of G' are three sided. The requirement that all faces of G have at least five boundary edges is equivalent to insisting that all vertices of G' have degree at least 5. The Four Color Problem now requires that we color the vertices of G' (using no more than four colors) in such a way that two vertices which are joined by an edge in G' have different colors.

This approach to the Four Color Problem as one involving vertex coloring is the one that most mathematicians have used recently. In the Appel and Haken proof, all results are in terms of this formulation.

References

1. Appel, K. & W. Haken. 1977. Every planar map is four colorable. Ill. J. Math. (September). **21**(3): 429–567.
2. Appel, K. & W. Haken. 1977. The solution of the four-color-map problem. Scientific American (October). pp. 108–121.
3. LaBrecque, M. 1970. The chromatics of cartography. The Sciences (April). **10**(4): 11–16.
4. Ore, O. 1967. The Four Color Problem. Academic Press, New York, N.Y.
5. Ore, O. & J. G. Stemple. 1970. Numerical calculations on the four-color problem. J. Comb. Theory (January). **8**(1): 65–78.
6. Steen, L. 1976. The four color theorem. Mathematics Magazine (September). **49**(4): 219–222.
7. Stromquist, W. 1975. The four-color theorem for small maps. J. Comb. Theory, Ser. B (December). **19**(3): 256–268.